GALACTIC
WHISPERS
CONVERSATIONS
WITH THE COSMOS

Ing. Hugo E. Belmonte González

D.L.: 4-1-1684-02

ISBN: 9798876308436
Seal: Independently published

ii

CONTENT

INTRODUCTION

The present new edition is the result of my passion for astronomy, which I took on as a hobby and in the amateur sphere, allowing me to particularly understand the immensity of distances and the enormous time that has passed since we don't know when (infinite), with the clarification that everything, absolutely everything moves in the universe, with absolute components (established by human beings), and relative ones such as with respect to the Sun in most cases) reaching in many cases the speed of light. The claim of all is to establish the age of the universe which for me is greater than everything that is mentioned because the existence of multiverses continues to be established and not just one universe as until recently. We can establish that five millennia or more or less represent almost nothing of the time that has passed since man inhabits the World, therefore, crudely speaking we are contemporaries of Jesus Christ and his 12 apostles, or Ptolemy and the great Greek sages or Newton and Einstein, in the future archaeologists will easily confuse us with them if our remains are not cremated. I dared to say the above because as we will show, what we are currently enjoying, basically the evolution of electronics has been developed by the most prominent of those we are calling our contemporaries.

When I was a child I think and remember, before I was 4 years old I used to listen to conversations between my parents with my uncles and their friends and what caught my attention the most is that they said that the World is a sphere and what I understood is that we were inside that sphere within which during the day we only saw the Sun and on some nights the Moon, crescent, full or waning with its characteristic shape which according to drawings from that time even had a little clown sitting on one of its ends and accompanied by little lights everywhere and that over time I thought they were the famous distant cities and that if one traveled by train one would arrive at them following the inner curvature of the globe, I even thought that with a good ladder it would be easy to get to the Moon by the direct route. In the city of La Paz Bolivia in general the sky is very clear or celestial especially in the winter. But in the midst of all this came the big question that worried me until I came of age and that was, what would happen if that shell (similar to inflated balls) that enclosed us was pierced, what was out there and that is what I imagined as a great void. After my school studies and the time it took me to get my degree, I resumed this hobby, especially one day at 5:00 pm when I saw that the Moon in front of my window was in a first quarter and due to the few clouds that are usually in our sky, I decided to check the next day at the same time what was happening with the Moon and it had moved to appear in the same place later and a little more grown, which led me to follow that movement the next day and indeed after the same time the Moon appeared fuller, I found that it does indeed rotate in the same direction as the World and apparently because the tangential speed of the World is greater it would seem and it is true that the Moon hides appearing to rotate in the opposite direction but it is the World that overtakes it and as it rotates in the direction of the World the next day it will appear with an angular displacement of approximately 12.19° . (The Moon rotates in the same direction as the World almost on the same axis with oscillations of up to 5°).

Trying to improve the second edition of this work and with the frequent press publications such as the bibliography used, they show that the universe is plagued with solar systems many of them with singular resemblance to ours, that is with 8, 9 10 or more planets among them already confirmed spinning around suns (Legitimate solar systems embedded in galactic clusters). Not all galaxies have the same characteristics as ours, known as the Milky Way, which for greater objectivity resembles the light foam of a hot coffee after stirring the sugar. In astronomy, answers to new curiosities continue to be found, which is why this edition is greatly enriched with them.

Our planet is immersed in the aforementioned galaxy with more than 100 billion suns (stars) and quite external to it, which in turn has allowed detecting at least other galaxies and several of them, of the same type or configuration as ours with approximately also 100 billion galaxies, which are rotating at abysmal distances. In those proportions there is the possibility of planets occupied by the three kingdoms of nature, that is, beings as human as ourselves, animals as varied as those we know and vegetation like that of our Amazon and our summits and highlands, even under the sea (water)) and surely a moon to regulate the movement of fluid waters. It is likely that many of these lands have living dinosaurs. (If several of them could be observed at the same time, the history of the evolution of our planet from its origins to the present day and beyond, that is, the future, would be very clear).

Another noteworthy particularity is that lengths are measured in terms of speed and thus the average density of suns within our galaxy is equivalent to a solar system within a cube with edges of 7 light years (travel for 7 years at the speed of light). Inconceivable that suns or stars collide, there are photographs of galaxies that are intersecting. Among the many billions of suns, some surely got out of control and collided with some other, but at distances that will not affect us even with a single atom. Supernovae are the result of stars colliding.

Consequently, for us the big problem is the time and space that have passed, are in progress and have yet to pass, to continually keep discovering what lies beyond, particularly contacting beings from other planets that apparently and as described in this publication, are very similar to us and the same, their animals and their plants clarifying in each case that they are not necessarily living in our era (relative to the time timed by our clocks), it is likely very outdated or very advanced, because the differences are astronomical numbers and because there are an immense number of planets, there must be twins to ours, the problem is locating them and contacting something that will not be achieved even with email. Because we must also assume that there are planets in formation and others that have ceased to have people, animals and plants, that is, they are dead. The evolution of life on World must be in different phases among the multitude of planets. For us, the bone remains from millions of years ago practically correspond to a certain period, so saying that the beings that inhabit the planet World for example three millennia belong to the same epoch on the timeline, however, surely Christ did not know the wonders of cybernetics, not even Don Simón Bolívar or Don Franz Tamayo. It turns out then that what we are savoring is very timely at this moment in our lives based on the work of contemporary ancestors.

Speaking of clocks, those of the World control the rotation of our planet around its own axis, around the Sun and finally around our galactic center, but the accuracy of this one does not have to be the same as that of other planets, since it is relative to their system. When we establish contact, they will have to let us know what kind of clocks they have and how we can make them compatible with millennia, centuries, years, months, weeks, days and hours with minutes and seconds. Another detail that we must not neglect is that the small is so tiny that somewhere I read that the day we can describe a large grain of rock, we will have come across the composition of the universe.

Finally, I wish that reading this document is accessible to all audiences, from school to university in all its specialties. We are not the only ones in the multiverse.

At some point it is necessary to talk about the infinitely small, perhaps mention the particle of God. This controls all the genetic information of each being including its link with our ancestors and surely with our successors, because family traits are normally easily identifiable. This particle must be part of the genetics of animals and plants, because seeds are exclusive to each species.

The biggest problem is that we are living in an age when research is endless and much more so with the help of IT that accelerates it, making the present collection of information obsolete in a short time.

Hugo Edmundo Belmonte González

1 QUESTIONS TO THE UNIVERSE

For many long years I have been asking myself a series of questions or curiosities, many of them answered daily through the press and television worldwide, about the experiments that are surely carried out inside the spacecraft that are launched into space with or without astronauts, both during the outward journey and when entering orbit or landing on the moon and then returning. What I heard and recorded very well is that the International Space Station, in an hour and a half of our clocks, makes one revolution around our planet. By the way, on World **we land**, and on the Moon we **touch down**, but the samples that are brought from our satellite and surely those that will be collected from the planets of our solar system, as well as meteorites, are practically rocks and material similar to that of our planet, so for greater convenience, we generalize the expression of **settling on the surface** of any of the planets, including exoplanets that belong to other suns, as landing and consequently the World should preferably be called by one of its other names, such as **World**, which unlike the other planets and satellites of our solar system has a large amount of water in a liquid state and air with water content in a vapor state on its surface mainly composed of the material called World that give it the characteristic of being a *blue planet*, which we know is fundamental to give us life, to both plants and animals and to humans who are the ones who will understand what we go on to question below: The layer known as the atmosphere allows airplanes to fly up to a certain altitude above the World because among the gases that compose it there is also oxygen that allows the combustion of the kerosene used by airplane turbines and logically between clouds that show the concentration of water in a vapor state. That is, the atmosphere with high and low tides (pressure changes) allows the lift of flying birds, airplanes, parachutes and paragliders.

Of all the planets in our solar system, the World is the only one with the capacity to admit and generate life, both to the animal kingdom and to the plant kingdom with beings like us who grasp all this marvel and enjoy it, except the mentally ill (madmen) who do not appreciate anything or very little and on the contrary generate discomfort as a negative component of life. If life is found in its other planets and satellites in our very small solar system, it will be at a microbial level, especially due to the lack of atmosphere and particularly oxygen and what is more water. Ultraviolet rays apparently even kill microbes.

But surely outside our system, where there are more than 100 billion solar systems with many suns very similar to our star called SUN. Through combinations, variations and permutations the possibilities of planets very similar to our planet that I prefer to call World are excessive, because there is World on all inhabited or uninhabited planets.

The large observatories including the external ones that orbit the World and their research teams have counted to date more than 500 planets around several suns among so many solar systems present in our galaxy, the known Milky Way.

The thing goes much further because more than 100 billion galaxies have also been detected with approximately another 100 billion suns each, clarifying that in no way is it mentioned that we are observing the infinite limits of the universe, if that happened, we would be seeing God. It also seems that there are many other immense *universes* that together receive the name of **multiverse**.

There is no doubt that the presence of human beings in all these universes is real. The regrettable thing is that we have not managed to make contact neither in the past, nor ever, nor even with those of our own galaxy. Perhaps dreamers have done so because the stories they spread are just dreams and we have no reason to believe them. The hieroglyphics found on World are the work of terrestrial ancestors of ours, no more than 2.5 billion years ago, times when human intelligence also had many limitations. To date, this intelligence continues to evolve, as evidenced by the racial groups and the predominant measurement of the intellectual quotient in certain races.

Some of the solar systems may have planets with moons and also with life and thus have easily established contact between two nearby planets, if they were going through moments of evolution like our present. For approximately three millennia (an extremely short time), we have had means of transport that are not horses, I mean cars, ships, airplanes and rockets into space, which with more sophisticated designs have managed to get out of our gravitational attraction field to take humans to the Moon just half a century ago. Voyager is a rocket that sends us images of space and apparently is already leaving our solar system. The images take more and more time to arrive to World as Voyager moves away. Remembering that human beings have been present on World for about 2.5 billion years and we are just trying to conquer space.

Just as the science of electronics manipulates atoms facilitating large procedures at the speed of light, as in computing, at some point they will manipulate the rays of light itself, perhaps finding or generating much finer and faster rays, because not even the speed of light allows us to shorten the immensity of spaces expressed in terms of the movement or journey of light rays, endless by the way since it radiates in absolutely all directions and for millions of light years arrives at our eyes especially on starry nights, with little or no light reflected by the Moon or the lights of cities which in any case harm night observations.

With good telescopes you can better appreciate the stars or suns that in the form of milk are seen with the naked eye on starry nights, including resolving nearby or distant stellar clusters which also contain millions of suns. The ones closest to the center of their galaxies must be very hot and therefore with too little probability of life, but this should not represent more than 60%, and counting only 20%, or less that have inhabited planets which is already more than enough.

Even more, for these planets to be similar to the World they must contain seas (salty or perhaps not), air, perhaps oil and a good moon, capable of causing tides and stirring the atmosphere (the clouds provide shade protecting the covered parts from the heat and when very loaded they precipitate rain, hail or snow). The other components will surely have them, that is, minerals in accordance with Mendeleev's table.

In summary, the World needs its snow-capped mountains, its rivers, rain showers to maintain vegetation with the fertility of the World, sustenance of animal life. All of the above allows waterfalls to be transformed into electrical energy and with it into many other forms.

Another limit is the ambient temperature not excessively hot above 40°C (104°F), nor excessively cold below -20°C (-4°F). The waters of its seas I imagine must also be salty which has to do with the electromagnetic lines that guide compasses and define the magnetic poles and due to the rotation, the geographic ones and the time control. In short, there are many things that may or may not have and see if with them life is feasible as in our Worldly paradise without Adam and Eve that would be added to the story of Snow White and the seven dwarves. Speaking of temperature, it seems to be very similar throughout the **multiverse**.

We must not forget that on the Moon it is not possible to light a match or make a kite or fly a kite take off, and even less to take off the astronaut suit with its tremendous diving suit and a good umbilical cord of oxygen.

We know that with altitude the atmospheric pressure decreases and that the gravitational attraction keeps the water on the World and on top the air adhered to the World and on the air of the sea, with the characteristic that this is invisible when there is no fog for us, but the birds and the planes fly feeding at the same time on the oxygen content and the balloons are inflated because we fill them with the air we breathe and that at sea level it presses with one atmosphere that is reducing with height, and with the exception that being less dense is what is precisely found above the sea and the World and does not go beyond for the gravity of the planet, then come the first questions:

If the balloons inflated with air from the terrestrial orb or better World traveled to the stratosphere inside a spacecraft and others theoretically outside the ship; surely in the first case the interior pressure would increase if the spacecraft do not replace the pressure of their interior air, they would grow but since the ships must be pressurized they could remain constant, instead those pulled out of the ship (impossible) would grow so much that at one point they would burst (abstracting from friction), but here another question arises and that is where the air that the balloon carried and the rubber itself that constituted it would end up, would they float in space or be attracted by the gravity of the World, at what height above sea level does it remain floating in space and with the impulse or inertia of the ship begin to orbit the World, or the Moon or the other planets and perhaps the Sun, these are questions for the astronauts.

Here something else arises again and that is, if the air particles remain together (compact) or if they disperse or expand at the level of isolated molecules in space (perhaps towards the void). It is better if the balloons are inflated with water, all the above can be more objective and in sight because the water could even be colored, to observe this spectacle (for which the ship would also have to stop unless by inertia they follow it). Another possibility is that they violently transform into energy. The vacuum has been mentioned which apparently is not so empty because light is transmitted through it, the force of universal attraction, television waves with light and sound, apparently the energy called black is what occupies the vacuum.

Regardless of the pressure, the gravitational force decreases with height to the point that liquids do not require containers and float compact at least inside the ships, as astronauts are seen floating inside them and to drink them, the astronauts absorb them with or without a straw, so we emphasize again, how will these balls of liquid behave outside the ship? Most likely the intense rays of the Sun will disperse them into molecules to atoms or less. What would happen if on a trip to the Moon a bucket full of water is deposited on its surface, will it evaporate immediately if it is on the face exposed to the Sun and on the contrary will remain frozen on the night side until the Sun comes out again on the lunar surface whose complete rotation is around the axis of the World in approximately 29.53 World days which comes to be one lunar day. (Trying to match that in the past it was established that every 30 or 31 days is a month on World). The water when evaporating will go from the bucket to constitute a minimum atmosphere on the surface of the Moon or will be absorbed by the dryness of its land. Will it be possible for astronauts to take space walks or walks on the Moon without their space suits and particularly without their diving suits? This is possible inside the ships, but when exiting outside there is no atmosphere for breathing and the Sun heats or burns with temperatures above the boiling point of water, which means that water boils and evaporates immediately, but the question is whether this has been verified in some way with the water itself brought from our planet? The faces in perfect shadow (where nothing is seen, not even what is covered by the shadow itself, that is, black) will be frozen. Oxygen is missing for breathing or simply to make fire.

While the Moon has Worldly materials, unfortunately it does not have an atmosphere or appreciable amount of water, which are vital elements, on the other hand since it does not rotate around its own axis but around the axis of the World, its days and nights they are very long, that is, almost two weeks in the shade that due to the lack of atmosphere is very cold (below -100°C (-148 °F)) and another two weeks with very scorching sunlight because it is above +100°C (212 °F), and on the other hand, wind is not known there and less it's refreshing breeze, nothing flies and a cigarette cannot be lit.

We must not fail to talk about water, the Spanish name for the molecules that make up the great fluid that at first covers 3/4 of the surface of the planet World.

Its predominantly liquid state at present corresponds to a certain position of our planet in relation to the Sun, both in terms of time and space, which causes the seas to present us with hot, cold water, frozen at the poles and clouds in the sky. This does not happen on the other planets satellites of our Sun because they do not have the same characteristics of ambient temperature, they are not at the same distance from the Sun as us. We are talking about life on World, the only planet in our very tiny solar system populated by living beings both within the water, in the atmosphere and belonging to the animal and plant kingdoms. At some point we will come to the conclusion that life is controlled by a tiny vibration that moves within the brain in the case of humans and animals and somewhere in plants and their seeds, completely controlling the nervous systems and all the organs necessary for life, this topic is for doctors.

The water molecule H_2O is known to all as the interconnection of two hydrogen atoms and one oxygen atom, so small that it is not possible to see it isolated with its components but as an imaginative resource, yes on a wet surface reflecting the rays of the Sun evaporation is observed reducing the wet spot little by little until it goes through a droplet of water and it evaporates which is verified because its reflection of sunlight is reduced, until the water disappears quickly showing the reflected ray smaller and smaller which has to be followed with the sight and by the presence of indication of drop of water which is the smallest dimension visible to our eye, what evaporated went into the atmosphere where millions of these molecules swarm.

Scientists and researchers manage to extract oxygen from mineral compounds and hydrogen itself to combine them and artificially produce water, which is almost impossible outside the World 's surface because you need to start a fire to melt the components by heat and without atmosphere it is not possible to light a cigarette or on the Moon or in space outside the shuttle ships.

Millions of years ago, several other planets in the solar system also naturally or spontaneously had the liquid element in proportion to the World's seas. It is possible that their orbits around the Sun at some point had radii similar to the current one of the World.

Let us imagine the Moon with water (Which can be recreated on a computer), up to an average level between its protrusions and depths mostly craters or volcanoes, the result would be that its land surfaces (*Moon component since it is also possible to land on it*) would be something like annular islands or volcano peaks probably with water in their craters and protruding from the water like semi-pointed islands.

2 THE ATMOSPHERE OR AIR OF THE WORLD

With average thicknesses on the order of 200 km (124 miles), it is a protection and encouragement of life that the planet has, especially with regard to the impact of meteoroids. It is fundamentally composed for thousands of years of a vital poison layer, composed of ozone. It serves as a shield to protect the planet against the harmful ultraviolet radiation from the Sun. The atmosphere, which is made up predominantly of oxygen, hydrogen, nitrogen, and anhydrides plus water particles that make up the humidity in the atmosphere, ends at a height between 20 and 40 km (12.4 to 24.8 miles) similar to magnetic field lines. This ozone layer, an allotropic state of oxygen, contains in its molecular unit 3 oxygen atoms (O_3) instead of the 2 of regular oxygen (O_2). Looking at graphs obtained from mathematical models in the northern hemisphere in 1979, it had close to 400 Dobson units (*a scale ranging from 180 to 400*). By 2007, it was down to 320 and in the southern hemisphere it went from 280 in 1979 down to less than 220 in 2007. Mexican experts predict a recovery of this important layer by 2050. They hold that this will be possible due to the success of the Montreal Protocol, through which 190 countries agreed to reduce or completely eliminate emissions of substances that affect the ozone layer. If this layer disappeared, the Sun's ultraviolet light would completely sterilize the World 's surface and annihilate all terrestrial life.

Moving to another topic and thinking about supplying water to satellites that don't have it in the future, it is probable that comets with the most elemental substance could be steered towards them since they are immense balls of water and other substances, frozen as they move away from the Sun, and to the contrary as they get closer, they emit a gaseous tail directed opposite to the solar wind, that is, perpendicular to the Sun's surface instead of behind as would be normal for tails. The detected presence of water at the poles of the Moon seems to be due to comets that impacted it, forming streams that reached the very dark regions near the poles. (If the water remained in shadow and without atmosphere, it froze and remains so if it does not receive sunlight to this day).

Water in its three states, as we know it in the world, is very unlikely outside of it. We are 148 million kilometers (92 million miles) from the Sun. The Moon is also at this average distance from the Sun, but does not rotate on its own axis. It didn't retain its water because it orbits around the World and not in an independent orbit, and even more so with one side always facing the World, meaning the radius of its centrifugal force goes from the center of the World at approximately 300,000 km (186,000 miles). Any water it may have had was ejected into space and perhaps some went to World.

The Moon is World 's partner, which is the center of rotation for both, as well as an element that plays with World 's water levels causing high and low tides, because it revolves around in approximately 29.53 days while World revolves around its axis completely in 24 hours. The atmosphere, with much more flexibility, also varies which is proven by barometers that can predict rain. In the southern hemisphere's winter and northern summer, the Moon and World are closest in their orbits to the Sun and 6 months later are farthest away. This is confirmed by the full moon which at one point looks small and at another quite large.

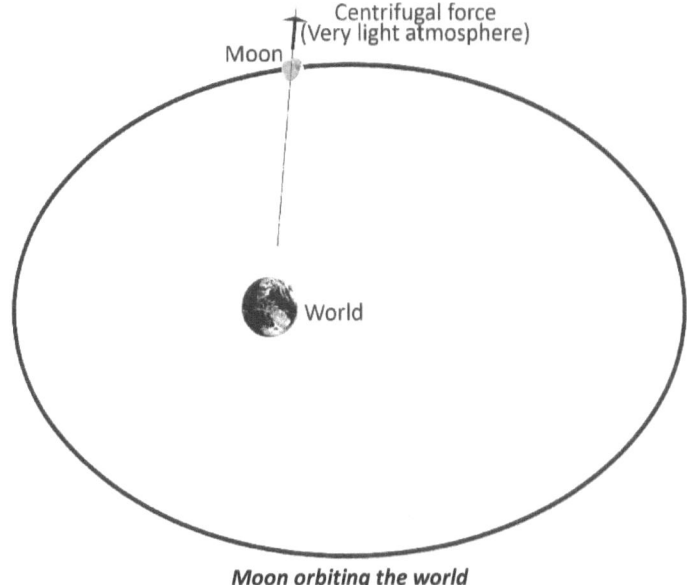

Moon orbiting the world

These conditions make it so that for the moment, researchers are looking among the 100 billion stars in our galaxy for similar proportional systems that allow them to state that there are millions of planets inhabited by beings very similar to Worldlings with their animals and plants. I would even dare to say that given the possibilities for combinations in the human genome that contains all the genetic information which has between 20,000 to 25,000 genes distributed in 23 chromosome pairs in the cell (the basic unit of life) and dealing with millions of inhabited planets with millions of people it is almost certain that there are natural clones of several people among us and perhaps multiple ones in different parts of the universe due to combinations of those chromosome pairs, of course very far from us. The cell is the smallest structure capable of carrying out the three vital functions on its own - nutrition, communication, reproduction. It is impossible to find our clones and converse with them. It is good to clarify here that mitosis is the division of somatic (non-sexual) cell nuclei, resulting in the exact division of genetic information. In any case, here on World we have grouped several races, that is, differences between humans because for example we distinguish white people with blonde hair, very dark black people with black, curly hair, Asians with slanted eyes, Latinos, Hindus different from all of them and with countless languages including different writing scripts. It is possible that with the advancement of technology some humans have been cloned here on World but due to religious reasons there is no public information.

The Sun's heat gets to World very well distributed thanks to the humidity from the seas and particularly the atmosphere with clouds that diminish the concentration of light which is reflected even by atoms allowing us to see objects including in the penumbra, impossible on the Moon because there the shadow is black without reflecting what it covers. From World during a new moon, we can clearly see the part illuminated by the Sun, but it is also possible to see very faintly that the unilluminated part shows its mountains because the light from World reflects about 5% thus illuminating it in reality.

The temperature ranges are such that they allow life, due to a slight inclination of World 's axis in relation to the plane of its orbit. In winter there are relative cold temperatures and in summer heat that can be combatted with cool water or fans. The equatorial region between the tropics is the one that best takes advantage of the heat and logically the polar regions have their frozen water. Animals are perfectly adapted to the region they inhabit; therefore, it is not advisable to change their habitat.

Speaking in general terms, the heat from the millions of suns plays very important roles because as in the case of hot water which has a lower density and rises in the atmosphere, and on the contrary cold water with greater density precipitates in the form of water, snow or hail, which is expected to be common in the case of exoplanets (current designation of planets outside our solar system).

Remembering that the average radius of World is 6,367.47 km (3,960 miles), the elevation differences above sea level have a maximum of 8.5 km (5.3 miles) (Himalayas 0.13%) and thus the altiplano in South America has an average elevation of 4 km (2.5 miles) and even though it is located in the tropics, the mountains become natural dams of frozen water known as mountains with perpetual snow. These protrusions are minimal and World is a very perfect sphere (although the slight flattening at the poles also makes it known as a peroid of revolution), currently in the process of melting because humans are deteriorating the ozone layer that caps off the atmosphere which originally had a thickness of 0.2 inches (5 mm) and has declined to 0.16 inches (4 mm). Consequently, protection from infrared rays has been reduced by 20% at the moment, increasingly damaging human skin. Most concerning is the thawing of mountain snow which are natural dams regulating river water permanence. If [the snow] dries up, drinking water will not be present on World. It is very important to replenish these reservoirs with artificial dams because in less than 50 years the snow-capped peaks will have disappeared and the freshwater running through the planet's vein-rivers that complement the hydrologic cycle, irrigating the planet's surface and allowing the development of all plant life and livestock hydration will be extinguished. This is very difficult to do with seawater which is very salty and would have to be pumped to said altitudes. Temperature variations cause evaporation of salt-free seawater and precipitation on land which when cooled transforms into snow and water in rivers that drain back to sea. (This is complementary to the four seasons). We see the Moon as a perfect sphere and surely from it we see World also as a larger celestial sphere, even more so with its seas and clouds leveling its surface.

On the other hand, the water coming down from tall mountains, if piped appropriately through pipes powers hydroelectric turbines generating energy to supply cities both in their nighttime lighting as well as powering multiple electrical systems, a large quantity of home appliances. This form of energy is reinforced by the direct use of solar rays, as well as petroleum sources that have been generated by the planet's geology. Photographs obtained from artificial satellites show World at night with a large number of bright spots in regions not covered by clouds which are city lights that in the case of coasts clarify for us the limits between the sea and land.

Life on the planet we inhabit is also possible thanks to water apparently moving on its own and in turn presenting itself in three different states (without taking into account that the atoms comprising it are tiny systems in motion similar to solar systems) due to the universal displacement of all the galaxies and each of their components. We can recall that we are all spinning at great speed around World 's axis, movement that is in reality also the endless source of energy that is transformed into various types used by humans. [World] in turn [spins] around the Sun which internally shifts in its galaxy which in turn spins around the galactic center and it along with the other galaxies at speeds exceeding the speed of light which allows one to speak of infinite distances and time. Regarding what was mentioned, our Milky Way has a pair of small satellite galaxies orbiting it with respect to the center of the group.

With certainty, the known elements of Mendeleev's periodic table constitute each and every one of the bodies that make up planets, suns, galaxies and galactic clusters with predominance of some minerals like silicon, iron, nickel or aluminum crystallized or melted forming magmas, with enormous temperature differences. Probably galactic centers are predominantly radioactive materials. Suns (stars) are made of hydrogen and helium which through their continuous mutation cause nuclear explosions (sun spots) in the manner of flames that transmit their heat and light throughout the universe and despite that, their life is and will be extremely prolonged.

Light rays play an important role because they allow us to see everything existing in the universe, since although irradiated by each sun they travel in absolutely all directions perpendicular to each one and it is of course possible to see them and observe them from any other distant solar system millions of light years away. One can also see the same stars whose light reaches us and we think may even be stars that ceased existing millions of light years ago, but we continue seeing them because their light continues traversing intergalactic distances which, we stress again, are unfathomable. The most impressive thing about light is that it bounces off different surfaces and breaks down allowing us to recognize object colors, textures, shapes, etc.

These rays not only conduct light but sound and even heat. As light breaks down into colors, it is possible by the refraction of rays emitted by some object to determine the predominance of components comprising the periodic table of elements - the case of suns and on the other hand, planets. In some surfaces it is reflected, as in the case of mirrors, which in the universe can reflect the hidden, extending light's path. It would seem entire galaxies can reflect in space because we appreciate them as comet nuclei without their tails (diffuse spots a bit different from actual stars).

What is interesting about astronomical research is that indeed thanks to waves the presence of planets has been detected particularly around other solar systems not only in our galaxy. It is very difficult to locate them because the light from their stellar luminary or sun outshines that reflected by the planets which unlike ours have no light of their own. With the exception of Mercury and Venus which are in front of World, we see all other planets especially at night completely illuminated by the Sun. Mercury and Venus exhibit phases similar to those of the Moon in their crescents or wanes and are seen as the evening or morning stars.

Something very important to highlight is that minerals melt at different temperatures. Consequently, in the location of our World, the predominant water is found in liquid state and to a lesser extent in gaseous and solid states. There is also a metal found in liquid state on World - mercury - which is normally extracted from cinnabar (a red mercuric sulfide mineral). Also, all alcohols and petroleum derivatives as well as gases like helium, oxygen, hydrogen and nitrogen. To separate gases from air, it is frozen and as the temperature rises different components are recovered individually. All this within the temperature ranges on World 's surface.

Having clarified this, hotter planets must be formed by similar substances but hotter along with other melted or liquid minerals. And vice versa - colder ones more solid with very few liquids and gases, all this on their outer face or surface because even inside World, as one penetrates into its interior or better yet its core, there are minerals permanently melted like iron and silicon which are fluid and which in the case of volcanoes are emitted along with other materials in the form of magma, with emissions of smoke and lava. Crystallized or non-crystallized silicon is the most predominant element in the universe.

3 THEY DISCOVER A PLANET SIMILAR TO THE WORLD

An international team of astronomers discovered outside the solar system the smallest planet most similar to World identified so far. The new planet has a mass five times greater than World 's and can be found 25,000 light years away in Ursa Major, orbiting around a small red star.

Its temperature is estimated to be around -220 degrees Celsius (-369 degrees Fahrenheit). The frigid temperatures make the chances of finding life unlikely.

The location of our solar system within our galaxy the Milky Way is not very close to its congested center of suns also called stars, it is rather a little on the outskirts and since the galaxy is quite flattened, it is possible for us to find or see other galaxies of very diverse shapes including stellar clusters as satellites of galaxies and deduce or infer the spiral shape known as analogous to some of them (see graphics on pages 26 and 27) therefore solar systems located closer to their galactic centers are assumed that chances of human life similar to us are minimal there, with the clarification that there are more than 100 billion galaxies (not including those of the multiverse) in existence, it turns out that with certainty there are many planets similar to ours with very similar characteristics because as we have seen even the Moon is essential complemented by the tilt of the planet for our life, taking the opportunity to comment about the 4 seasons per year with their peaks as equinoxes and solstices that as stations begin a month and a half before each peak and end a month and a half later, because it is not that the seasons are advanced as is frequently heard especially with winter that in mid-May in the southern hemisphere it is said to be advancing, what happens is that it starts to take effect around May 5, reaching its peak on June 21, the date known as the winter onset and it is rather the winter solstice. Winter ending our winter on August 5 each year. *(See following graph)*

The ecliptic or annual path of the Sun in the celestial sphere as seen from World forms an angle of 23°27 ', to which the seasons of the year and also the eclipses are due.

As a consequence of the above it can be seen that it is time to make some corrections in the prorating of the days of the year, the naming of the months, modifying from the first day of the new year, which would result in modifying all movable holidays from religious customs.

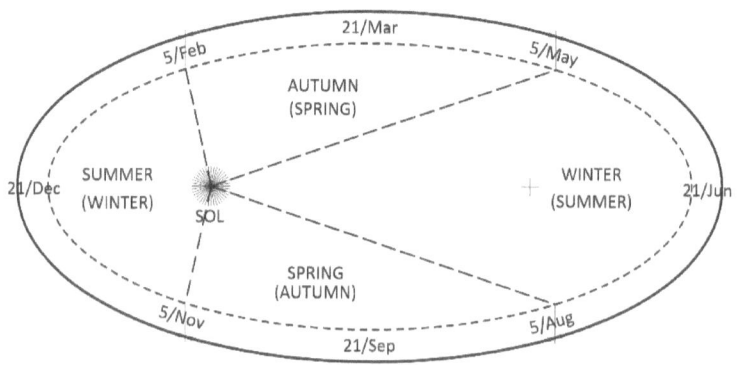

Seasons of the year for the Southern Hemisphere

(In parentheses for the Northern)

The number of days in the year is 365 and every four years 366 with other corrections and adjustments at certain times. What should be done is keep the number of months of the year as 12 (number divisible by 2, 3, 4 and 6) but with more rational denominations than the current ones as would be October which means 8 is the tenth month of the year, November 9 is the 11th, December 10 and is the 12th.

As the seasons of the year are marked every 3 months, it is necessary to group them as such (taking into account the known differences between the Northern and Southern hemispheres).

As 365 is not a multiple of 4, they cannot be groups of 91 days because that would only cover 394 days and there have to be exactly 365 days with one of 366 variables. The solution is: Three months of 91 days, the next three of 92 days, then another three of 91 days and finally the variable group with 91 days and every four years with 92 days.

In order to frame the four seasons of the year it is advisable that the first day of the year be February 5 of the current calendar or the one defined by expert astronomers.

The names would have to be defined in a universal agreement, but for reference, the days, months and seasons of the year would be as follows:

First month with	31 days	First season (Fall/Spring)	91 days
Second month with	30 days		
Third month with	30 days		
Fourth month with	31 days	Second season (Winter/Summer)	92 days
Fifth month with	30 days		
Sixth month with	31 days		
Seventh month with	31 days	Third season (Spring/Fall)	91 days
Eight month with	30 days		
Ninth month	30 days		

Tenth month with	31 days	Fourth season	91 o 92
Eleventh month with	30 days	(Summer/Winter)	days
Twelfth month with	30 o 31 days		

That is, it covers as always: 91 + 92 + 91 + 91 (92) = 365 (366) days per year.

The weeks would continue with 7 days and the designation of these could be new and with common names in different languages. As a global result of this proposal the monthly salaries would be more uniform.

The drawback that arises is the identification of people and legal documents registered with the current system that will have to be seen if it is maintained in parallel and everything new enters the new system for which it will have to be seen if with the help of computing it can be quickly updated, which is very likely. It would seem that the age is modified, but not, what is modified is only the date of birthdays. Up to here this interesting parenthesis.

Returning to astronomy we also see that the Moon is not suitable for life because there is virtually no water in it and less air or atmosphere which is what distributes sunlight evenly so its face that receives sunlight is over 100°C (212°F) and the shadow side at -100°C (-148°F). Without a spacesuit including the helmet and its oxygen tube, it is impossible to have walks in shirtsleeves on its surface and least smoking a cigarette or cigar. Another detail is that both night and day have a duration of about two weeks each.

It is then here that it becomes necessary to describe in more detail the behavior of water in planets that contain it and are inhabited by thinking beings, that is, intelligent and also animals including microbes and all kinds of plants. Microbes survive in large temperature ranges particularly below 200°C (392°F) and at more than 4000 m.s.n.m. (13,000 feet) above sea level.

Between 32 and 212 ° F (0 and 100 ° C) water at sea level is liquid, cold towards 32 ° F (0 ° C) and hot towards 212 ° F (100 ° C). Due to its fluidity, it is possible to send it from its dams to cities through pipes thanks to the communicating vessel system and pressure heights, to all homes in towns for various uses and then evacuate it by the sewer system and after treatment to the rivers and thus returning up to the seas from which, again by evaporation (already without the sea salt) they reach the atmosphere returning to replenish the dams. This is a privilege enjoyed by almost all cities in the world and should be similar in exoplanet siblings with inhabitants. Human life apparently develops under these conditions and not outside this range.

The incredible thing about water is that all the properties especially piped water allows it to reach its destinations with enough pressure, as is the case with showers, sinks, tubs, kitchens that use the liquid element treated, purified and clean water for human hygiene, feeding, clothing, etc. Here one could describe in detail the multiple uses that water is given (vehicle washing, floor cleaning, watering gardens and crops, etc.). Its supply is equally to all houses and buildings in large and small cities. For very tall buildings, water tanks are used where water from the network is stored and then with regulating hydrofoils it is pumped to elevated tanks. Petroleum gas is transported in heavy cylinders and modernly by pipelines to private homes.

In other fields, we will only mention a few such as the fact that the human body is made up of a large percentage of water (between 70 and 75%) and naturally animals and vegetables especially fruits. Food for human beings is also prepared particularly based on water (soups, soft drinks, etc.)

The first locomotives and steamers operated with water, on the other hand, engines are cooled by recirculating water in radiators.

Pools complement human recreation, who do not live if there is no water, they also like to submerge in it and even dive. (With all the more reason on the beaches surrounding the sea).

If valuable minerals were found on the surface of the Moon or in exoplanets, they could be exploited with mechanical means, that is robots controlled and energized by sunlight from the planet and accumulate the fruit of their work in one place for ships to collect it later, which unfortunately would make it so expensive that these operations are not justified, nor if oil were found since its energy cannot be harnessed for lack of oxygen. In summary, we have to let the Moon continue to fulfill its function by providing us with full moon and new moon nights and in turn causing tides and humid storms in the environment, and at most install bases to continue exploring space and since it has one face fixed, fixed antennas for electronic order repetitions can be left, suitable for its movement around World. It is possible to see the movement of the clouds and the icebergs of the polar areas more comfortably than from the artificial satellites (without discarding air, ground and sea traffic control).

Mars on the other hand, rotates on its axis in 24.5 hours that is to say its days are comparable to those of World, but its year is almost double and also has a strong inclination with respect to the plane of rotation around the Sun, about 25 ° and its distance is also almost double, that is much colder. The mountains and valleys on Mars are very small or not very pronounced in relation to World. Its sphere is apparently smoother and it has two satellites made up of gigantic irregularly shaped rocks with an average size of 12 km. If a human were to land on these and try to jump, they would surely jump off into space, becoming a new satellite of Mars if they are not attracted back to its surface.

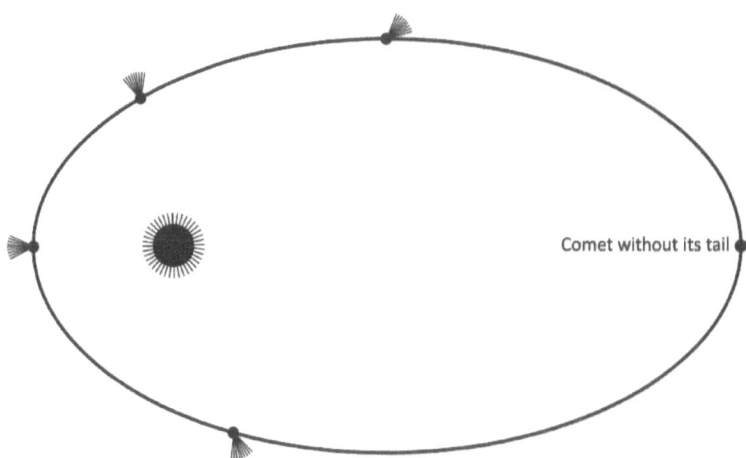

Comet without its tail

Comet in various positions of its trajectory

As a curiosity, the presence and approximate quantity of atoms of some elements from the periodic table valid for the entire universe per unit of volume beyond the stratosphere, that is to say in the vacuum (which from my point of view is not so empty), have especially been detected, particularly those with the fewest protons and electrons. This is because it is known that their presence, despite being very small, represents an obstacle to simply seeing beyond our solar system and even more so beyond our galaxy, due to the enormous distances. The small number of atoms accumulated is too high and causes interference in astronomical observations.

As astronautics progresses, ships that will rescue materials of all kinds are being sent more frequently, especially from comets or at least from their tails, which as is known go in the opposite direction to the location of the Sun, and not precisely behind the direction of travel, especially when the comet is returning, because the solar rays direct its tail in the opposite direction to the travel of the nucleus, that is, towards the emission of sunlight (known as the solar wind)

On the other hand, is space junk (ships or objects sent from World) becoming dangerous when re-entering the planet's surface, because while they do not return to World they occupy enough space so as not to damage anything, because in terms of the dimensions of the World's surface they occupy much larger spaces due to the radius of curvature of said surfaces being equivalent to the radius of World, plus the altitude above sea level which is greater than 200 km (124 mi) reaching 400 km (249 mi), which generates a much larger surface, where the junk in question will be widely scattered.

Meditating on gravitational force, let us imagine that humans manage to drill through World with a large enough diameter to allow the free circulation of bodies such as that of humans themselves (an absolutely ideal fact), and that it passes through to the antipode, that is from the nadir of a certain zenith that can be from Bolivia to the Philippines, crossing the magma core and all the component layers. It would turn out that if an object is thrown into this drilling on one side, it would come very close to exiting on the other side, but would not come out, rather it would fall back inside without managing to exit on the side from which it was thrown. That is to say, a pendulum movement would occur until the body was stabilized and trapped in the center of the planet. This is pure imagination since not even the companies that drill into the ground in search of oil or gas have reached 15 km (9 mi) in depth and with drill bits less than 0.5 m (1.6 ft) in diameter. And what they do drill, they do not necessarily drill vertically, but rather in case they encounter very hard layers they avoid them by curving their trajectory all retransmitted by means of computerized information for their control on the surface.

As for almost horizontal tunnels through mountains, enough of these are built because they are superficial and this is how the European continent has been crossed to the island of England under the English Channel with vehicle, railway and pipeline passages that necessarily have very powerful ventilation, lighting, water evacuation and ventilation systems because they are very long sections.

Evolution in the field of computerized encryption has allowed us to see large details of the topography of World directly with Google software, with streets and even vehicles, and it gives us very precise geographic coordinates in 3 Cartesian axes. Global Positioning Systems (GPS) are also common, which make it possible to plot the trajectories followed by any type of vehicle on plans and make it possible to return along the same route.

Astronauts surely handle high precision GPS; they also carry telescope-type cameras with large magnification range lenses. Consequently, they have achieved photographs not only of World, but also of details such as complete cities, with both daytime and nighttime shots. They surely do not reveal them to us, but with restrictions and limitations because evidently it is private property of the countries that invest a lot of money in their research. They must also continue to receive secret information from vehicles sent to the surface of the Moon and Mars. It is likely that these continue to move and extract samples or send photographs and television images that travel through the so-called vacuum which, I repeat, should not be so empty because rays of all kinds circulate through it thanks to which there is light and sound with which we see the closest solar systems to ours, that is within our galaxy known as the Milky Way, and perhaps we observe some other galaxies that we often appreciate as simple stars.

As is known, distances here are spoken of in terms of simple years, thousands and even millions of light years, which also makes one think that the speed of light (constant) probably becomes variable when certain enormous (inconceivable) dimensions or limits are exceeded. And if it comes bouncing off surfaces such as galactic sets, its straight trajectory would be broken and therefore longer, constituting all those reflections in true astronomical scale mirrors.

Rays of light (pure energy) possibly their range is even greater than the list of rays discovered on World. Perhaps many of them have infinitely smaller (or larger) frequencies than those of the known rays and communicate images much faster than light rays that allow us to see colors, due to their refraction, and thanks to that we appreciate the differences between objects and their three dimensions.

In any case, both the known and unknown rays move through the smallest or largest in every sense and this is how signals are being emitted from this planet, especially sound ones to see if distant planets capture these signals from some distant planet and try to communicate with our observatories. Because given the number of planets in the universe, there are surely living beings whose bodily components contain the known genes and their combinations (genome), insisting that there is life of beings as human as us and very similar, and likewise animals and plants. It is likely that on some of them there are even dinosaurs today which in the case of World gave rise to the existence of oil (underground fluid) which is currently used for combustion to power the machines and vehicles designed by man himself. In conclusion, there must be planets like ours with land and water and consequently with plant, animal and human life, and naturally with equal minerals to those we know. It is logical that if those beings communicate, they will do so with languages or languages quite different from the several we know on our planet and much more so with writing from papyrus to typographic writing or some other procedure they know. So the signals that are received or would have already been received are very difficult to respond to due to both the time it takes for communication and the language barrier. As a result, the maximum will be to try to respond with signals that so far take too long to reach their destination and with highly variable directions because everything moves in the universe, from atoms (their protons, electrons), galactic sets and with such precision, that the human being has had to build very expensive clocks that today with electronics have been perfected and serve to control the rotational movement of World around its own axis and around the Sun as well as the location of any bright point observed in the sky. Units to control time or rotational displacement must be different on planets in other systems and in some cases, they must have solar clocks or much better than ours because perhaps they have time advantages over us of greater or lesser magnitude. I dare say from galactic sets so far apart from each other. In the world, more than 200 languages are spoken as well as a similar number of different dialects, but we have the means for our intercommunication. It should be mentioned that translation of documents is already processed by software. The classification for evaluating the number of languages and dialects in the Worldish very complex. Not to mention the writing systems which are so diverse. Electronic dictionaries cover far more than the famous encyclopedias of the past (including films, photos and music or sound).

As can be seen, this subject gives rise to dreams that are too deep and which will be expanded throughout this work. As can be imagined, one begins to move away from questions to astronauts at astronomical distances, which is why it is very difficult to choose the title of this present work. I hope the one chosen is the most successful, because the telescope I have is amateur level and my knowledge comes from television and the press.

Returning to our solar system, there is talk of investigations searching for water both on the Moon and on Mars (where a robot has just landed for this purpose) and perhaps on one of Jupiter's satellites. The goal is to justify at least the existence of plant life (mosses, etc.) but this must be minimal. Now we have to ask, has water been taken to the Moon and has the way to conserve it been foreseen in order to achieve seed germination with lunar soil at 300,000 km from World? Possibly it will be necessary to additionally take natural or artificial fertilizer and protect it with tents providing shade, not necessarily solar ones, which retain ambient moisture.

If there is natural water conserved for thousands of years on the Moon or Mars, it has to be deposited under their surface or in some perfectly shaded cavern (where nothing can be seen because this satellite and that planet do not have atmosphere) and in the form of ice at extremely cold temperatures, as is known. Because if exposed to sunlight it would violently start boiling at high temperatures. Have the NASA astronauts surely taken thermometers for this type of survey and in that case at what temperature would water boil on the Moon? Because apparently since there is no atmosphere, the pressure must be bordering on zero and together with its small gravitational attraction bodies are very lightweight.

In the city of La Paz, Bolivia at almost 4000 m.s.n.m. (13,123 feet) above sea level, atmospheric pressure is around 0.67 atmosphere, the oxygen quantity is less and during winter, if there is a clear sky the Sun heats a lot and the dry cold is felt in the shade (it is not necessary to dress as warmly as in humid climates). This commentary is to relate what is imagined for the Moon since it makes us feel a bit of what must happen if there is no atmosphere. Another detail is that engine performance is lower.

The Moon does not offer favorable conditions for humans to settle there despite having practically the same distance from Worldto the Sun (~148 million km). What has been described above clearly shows that survival on the Moon, as free and sovereign as it is on World, is impossible except with bulky, uncomfortable space suits and helmets with an umbilical cord-type cable.

It is surely planned to build on the Moon, hermetic tents and resistant to solar rays into which air and water will be introduced, which will have to be generated by transformation of atoms or molecules, which seems impossible. Exiting these tents will be like going on spacewalks and let's not forget that gravity is so low that if you jump 1 m high on World, on the Moon this jump would be about 6 m or more, which suggests that building structures (because weight is a function of gravitational attraction) should be easier due to their lower self weight, as is the case of the longest bridges built by man on World which are suspension bridges and the longest central free span reaches 2.5 km. On the Moon for the same mass the weight is much less and consequently spans greater than 10 km can be achieved. (The problem is the raw material, fundamentally including water, so it will be necessary to transfer the necessary materials from here, since producing or extracting steel, wood, or concrete in large quantities will not be so easy). Moreover, since the Moon has no rivers, lakes or seas, its construction would only serve to cross deep gorges and grade separations. And this is when roads are built to interconnect areas conquered by man through astronauts who will continue as such for a long time. Having no atmosphere, not even lighting fires for smelting or other things is possible on the Moon. What real estate agents can do is make a fortune selling lunar territories to the unwary.

In summary, the Moon is a dry dirt ball near World which does not rotate on its own axis because it rotates around World's axis in approximately 29.53 of our days. From here it is observed in almost weekly phases and it does not have magnetic poles coinciding with its geographic poles, so orienting oneself with compasses is not easy. The Moon oscillates allowing 5% more of its exposed surface to be seen from World.

It is interesting to comment that in the United States, when the full moon comes out a second time in the same month, it is called a Blue Moon (The first moon between the 1st and 2nd of the month and the second between the 29th, 30th and 31st). That is 29.5 days later.

It is good to remember that there are two planets in our solar system that are closer to the Sun before World, and can eclipse the Sun without shading World due to their small size. They are hotter due to their proximity to the Sun and due to other conditions, however they are not suitable for life. Rather, we comment that to our view they have lunar-type phases and so when Venus is passing behind the Sun, we have a full Venus but the Sun does not allow us to appreciate it because it is also further from World. It presents its greatest luminosity (as the most visible planet) when it is a morning star crescent or an evening star waning crescent. (In ancient times they thought they were two different stars) and it is only Venus. Then Mercury is the other with phases but its greater proximity to the Sun and its smaller size makes locating it in space even more difficult. But by tracking them with computational electronic programs that show us stars on the screen, as planets, comets, etc. in their position at the time in question, if we see them with the naked eye in the horizon at certain hours before and after sunrise or sunset. Another possibility is that both planets can be seen in fairly close conjunction.

Galileo Galilei predicted the existence of Mercury, but he never managed to see it either with the naked eye or with lenses. It was based on Titius' law, improved by Bode, which allowed establishing its distance to the Sun with good approximation, including the planets known at that time. The law establishes that the relative distances follow the progression 4, 7, 10 (World), 16, 28, 52, 100, 196; This series is deduced by adding 4 to each number in the series 0, 3, 6, 12, 24... (3n, with n=0, 1, 2, 3, ...). Astronomers today believe this relationship is a pure mathematical coincidence. However, thanks to it the tiny asteroids between Mars and Jupiter were also found, Pallas being the largest one, similar to Mars' two satellites.

The simplest expression of Bode's law is:

$$d = \frac{(n-1) \times 3 + 4}{10}$$

Where:

d = distance proportional to 1 (unit) that corresponds to the World

n = Number of the planet.

If the World is number 3 it results: d = 1

The data approximate to reality are:

0.39	0.72	1.00	1.52	2.80	5.20	9.54	19.20	30.0
Mer	Ven	Ear	Mar	Ast	Jup	Sat	Ura	Nep

Surely professional astronomers have taken excellent photos of the solar eclipses of both Venus (which repeats every 112 years) and Mercury, which is something very beautiful to observe since the size ratio of both can be appreciated. (They are shown as round black dots on the Sun, and are even confused with sunspots). It is important to remember that the Sun cannot be observed directly, much less with lenses, it must be projected.

Very little Is known about Mercury, whereas it is known that Venus is made up of carbon dioxide without further details, and the fact is that the refraction of light is excessive in its atmosphere in which it is not possible to see at a distance (everything curves) or worse, not even the mountains are seen as we see those on World because they cover above the visual angle.

The precision with which the planets and naturally the entire universe move led to establishing formulas that later became those that gave fame to sages such as Tycho Brahe (1546-1601) Swedish Dane, who cataloged more than 700 stars that served so that based on this research those cited below in chronological order have taken the first major steps in the field of astronomy and particularly physics. He was the predecessor of Johannes Kepler (1571-1630) (known for his three laws) who initially worked as Tycho's assistant, then came Galileo Galilei (1564-1642) (invented the first telescope) and later Isaac Newton (1642-1727) (published the law of universal gravitation). Bode is also mentioned, who formulated an equation for the inter-separation of planets quite accurately, and this is how the presence of asteroids (planetoids) between Mars and Jupiter could be established, and it appears to be a planet that disintegrated in its formation process (it becomes a ring around the Sun, like Saturn's rings) and smaller ones in some other planet.

It is worth remembering Kepler's three laws:

1st Law - Planets orbit the Sun in elliptical orbits. The Sun occupies one of the foci of the ellipse.

2nd Law - Can be expressed as:

The areas swept by the segment joining the Sun and the planet (radius vector) are proportional to the times taken to trace them. This law implies that the radius vector sweeps equal areas in equal times. This implies that the orbital velocity is variable along the path of the celestial body, being maximum at perihelion and minimum at aphelion (the velocity of the celestial body would be constant if the orbit were a perfect circle). For example, World travels at 30.75 km/sec (19.12 miles/sec) at perihelion and drops to 28.76 km/sec (17.87 miles/sec) at aphelion.

3rd Law - The squares of the orbital periods p are proportional to the cubes of the semi-major axes a of the ellipse.

$$p^2 = K * a^3$$

On the other hand, Newton, based on previous studies, established that the force with which two bodies attract each other is proportional to the product of their masses and inversely proportional to the square of their distance.

Einstein's theory improved?

Some astronomers have refined Einstein's universal theory of gravitation, creating a "simple" theory that could solve a dark mystery that has confused astrophysicists for three quarters of a century.

The new law of gravity developed by Dr. Hong Sheng Zhao from the University of St Andrews and Dr. Bernoit Famaey from the Free University of Brussels, tries to demonstrate whether Einstein's theory was in fact correct and whether the astronomical mystery of "Dark Matter" really exists. The formula developed by both suggests that the intensity of gravity decreases more slowly with distance than in Einstein's theory, where it does so with the inverse of the square of the distance between the bodies, also subtly changing from solar systems to galaxies and from these to the universe as a whole.

The physical theories of gravity were first developed by Isaac Newton in 1687, and perfected by the general theory of relativity by Albert Einstein in 1905 and later years to take into account the curvature of light. Although it is the earliest known theory to man, gravity remains a mystery with aspects and conclusions still unconfirmed by astronomical observations in space.

The "*problem*" with Newton's and Einstein's golden laws is that although they work very well in the World, they do not explain the movement of the stars in the galaxies and the curvature of light with sufficient precision. In galaxies, the stars spin rapidly around a central point, held in orbit by the gravitational attraction of the rest of the galactic matter. However, astronomers found that these moved too quickly to be held by their mutual gravity; the fact that gravity was not enough to keep the stars bound together in the form of a galaxy should lead to the stars shooting off in endless directions!

Jupiter, being the largest planet in the solar system, exerts great attraction and thus attracts a good number of asteroids that approach it and are called Trojans and in many cases are surely captured by it. There are also some close to Mars and others approach the World and several of them pass very close to it, there are also sand grains that are seen as shooting stars because when penetrating the atmosphere, they become incandescent. In short, Jupiter has protected and will continue to protect the World from meteoroids. From here, approaching asteroids are controlled as much as possible and thus the one designated AA29, which approaches us every 95 years, its last visit was verified in 2002 and the passage of an asteroid the size of a 45 m high building (148 feet) was scheduled for 2013 at a distance of 27,000 km (16,777 miles) from World.

The astronauts who travel around the world, preferably over the equator, circle the world in an hour and a half, consequently, their biological clock (physiological system that allows organisms to live in harmony with the rhythms of nature) must have particular disorders linked to sleep, eating times and needs. Here we will check the speed of movement around the World with the help of the following graph: In which we verify the perimeter of the World at sea level and also at 400 km (249 miles) which is the flight altitude of the ships that circle the World and as we know that the time absorbed in each lap is an hour and a half, it is possible to quantify the relative speed at which the ship circulates, which is certainly absolutely greater than that of our airplanes.

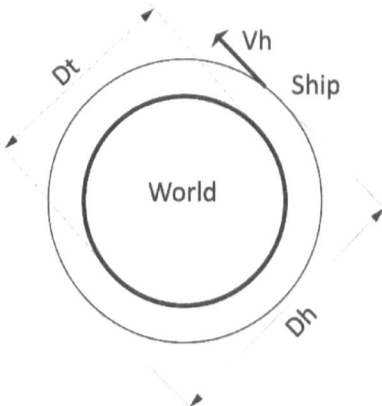

Orbit Above the World (Without Scale)

Adding 400 km (249 miles) in orbital height, its diameter will be 15,354.94 km (9544 miles) with a perimeter of 42,521.27 km (26,451 miles) and if the ship travels that orbit in 1.5 hours. Its speed will be 28,347.51 km/h (17,613 mph). Practically 7.88 km/sec (4.9 mi/sec) almost imperceptible compared to the speed of light, the ratio is 0.00002625. To travel to the closest exoplanet, you have to do it at the speed of light for more than 42 years.

On the other hand, the first asteroids that were detected and that are the largest (the size of cities with about 100,000 inhabitants) were designated as Pallas, Vesta, Juno and Ceres and others, perhaps compatible with the size of the two satellites of Mars, which are not spherical since they are deformed rocks, which may have been captured from the ring (indescribable problems occurred over the course of millions of years). As is known to all, for half a century the World has ceased to have only one satellite, because even man went out to orbit it, there are thousands of tiny artificial satellites and some are even considered to have passed into space junk status, in a few cases having re-entered the planet. Several satellites are stationary (used in telecommunications), that is, they rotate at the same angular velocity as the World and others orbit following meridians and/or parallels or perhaps in a retrograde or opposite direction to the rotation of the World and the solar system, what they do have to have is tangential velocity to counteract gravitational attraction that is, they have centripetal force that is compensated by the centrifugal force, similar to a stone that is made to spin tied to a string, it pulls to the point that if spun excessively, it breaks the cord and exits tangentially. There are temporary satellites where astronauts leave the space shuttles and take walks always tied to the mothership and also with a kind of gun so that by the principle of action and reaction they can move around and return somewhat comfortably, otherwise if they become detached from the ship they can float in space and here comes the question, will they be attracted to the Sun or to the World, or would they remain as satellites spinning around us or finally disintegrate? The known height for this is approximately 200 km (124 miles) from the World's surface, and there the World's attraction no longer applies (this attraction persists, but is negligible because it allows exiting into space) by the components of rotation, and the atmosphere ceases to exist and a vacuum appears that is full of only energy in the form of ultra-fine rays that are absolutely invisible and that allow the circulation or transmission of light and sound. What happens if astronauts kick a soccer ball into space, where would they score the goal? or if the solid rubber ball perhaps they will be the ones kicked off, maybe both in opposite directions.

environmental conditions of the World restored, so it is possible to take off the suits, especially the space suits and they must converse as on firm ground, but if they go out like this to the outside of their ship with oxygen balloons for breathing, will it be possible to converse and without emitting sound or will they do so via cell phones, will it sound louder, softer or not and if they emit sounds by bells, horns or other known means, will the same thing happen, apparently this is very complex to experience. The probable answer is yes, since spaceships transmit sound waves to the World similar to those of one or more stones thrown into water television signals, sound signals, etc., this means that these emissions also travel in the so-called vacuum and in very straight directions which increasingly makes us think that the vacuum itself does not exist as such because the waves go in all directions and are something like what we see in practice on the surface of the water on our planet, very clear, if we insert a saber into clear water it seems crooked or rather broken by the refraction of light, but if we dive into the water with the saber, we will see it without breaks, that is, objects and their colors are seen both in the air and in the water, as for sound it is clear in the air and too low in the water (perhaps fish perceive it very clearly), but now the vacuum comes in which is probably pure vibrant energy to transmit all this and at what speed? (Will it be the speed of light or the speed of sound?).

Within our solar system, galaxy and universal in general, light travels at a constant speed established by Albert Einstein and rounded to 300,000 km/sec (186,000 mi/sec). (Equivalent to its displacement from the Moon to the World in 1 second), it would have been interesting that in the afterlife it would increase in proportion to the immense distances, which would be very advantageous because it would shorten real distances, but this idea will surely take a very long time to be solved because to reach the sun closest to our Sun, that is, Alpha Centauri, with one of its satellites (another sun) called Proxima Centauri, a trip at the speed of light for 4.3 light years is required, that is, it will be necessary to go from mass to energy which is the form of high speed mobilization in space where the average density of solar systems corresponds to one sun (star) within a cube with edges of 7 light years, therefore it is admissible that some galaxies are even intersecting or intertwined or superimposed without danger of collision between their suns, nor their planets.

All this suggests that for the moment many, many years will go by before man alive can reach the distant planets of our system, because otherwise they have to transform themselves in some form of energy for the trip and once on other planets inside and outside our galaxy they can retransform into matter again (not necessarily) and thus be able to converse (by signs) with the inhabitants who surely live out there and who must be as many as on World, that is millions, so that the combination of the components of each human genome can now repeat themselves, in the past or in the future and if so we must have copies of ourselves very far away (something like cloning), also land and water must be in quantities grouped into larger or smaller dimensions in other solar systems called brothers of ours and with internal satellites such as the Moon regulating tides and helping distribute water over the planet's surface and even generating eclipses.

Water has the particularity of being the liquid element which by influence of the rays of light that mobilize heat, passes from solid to liquid and to steam. Without losing sight of the fact that there are liquid metals such as mercury and gases in the atmosphere, but the role of water is very objective of the influence of energy and matter, because water irrigates fields, freezes the top of the highest mountains and concentrates in the oceans with underwater currents due to thermal differences, evaporates and forms the known water cycle that in turn moves in terms of the four seasons of the year on the surface of the World. The movement of water allows the generation of energy, the erosion of the land and consequently the transformation of the land surface and on the other hand the watering of the forests.

Among the twin exoplanets to ours, there is also the probability that its surface covered by salt or non-salt water is in greater proportion than that of the World and even total, consequently in this last case, if there is human life, it has to be underwater, that is, there are necessarily great differences with the life of humans on World since there is not even the possibility of making fire in the water that is, they have no smelters, and logically they have no vehicles, no combustion engines. As for their clothing, perhaps they do not require it because instead of body hair they have scales and most worryingly they must be great swimmers so their fingers will have membranes to facilitate this typical displacement of frogs and ducks. so, it is not worth trying to compatibilized them with human beings, so they are excluded from the comparisons we have been making. In the too distant future, they will have to be included if they truly exist and it is proven that they are beings similar to us, it would be a new race and surely with varieties.

Once again stepping on firm ground on the World, it can be said that nature is so large and so well planned that the water cycles obey a Lunar Eros, that is, every 29.53 years the eclipses repeat themselves in the different areas of the World and also the variations in the tides and in the flows that circulate through the rivers, therefore the hydrological controls to project bridges over rivers must be at least 30 years and better if these controls are 100 years so as not to run the risk that the waters exceed the superstructures and serious accidents occur, noting that not all works of art have been executed taking this forecast into account. The alteration of the ozone layer can translate into overheating that can interfere with what has been mentioned.

It is also good to remember that water is naturally dammed up in the highest mountains where temperatures are below 0°C (32°F) giving rise to the eternal snows that constitute natural dams that regulate the currents of water, so if now having man reduced the layer Ozone discovered by Van Halen and expressed as 5 mm (0.2 inches) and currently reduced to 4 mm (0.16 inches) allowing greater filtration of ultraviolet rays that not only harm people's skin, but prevent the conservation of snow and thus deplete the large reservoirs of water that should be restored with artificial dams, because in a very short term, perhaps a century or less, they may cease to maintain snowy summits of high mountains that were called eternal snows, and consequently only the rains would fill the rivers when it rains in certain areas and regions and in reality so variable and as has been indicated its repetition is in periods of 30 years with their maximums and minimums, in between. In the city of La Paz Bolivia there was a very suitable snowy summit for snow skiing, competitions were held there, but to date the Chacaltaya snow-capped mountain (in La Paz, Bolivia) has lost its beautiful whiteness and several other snow-capped mountains show the progressive reduction of their eternal snow.

On the other hand, in cities, as they grow and their streets become paved, it turns out that the runoff caused by clouds in their rain precipitation is greater because less water is consumed, therefore it gives the sensation that recent rains are of greater intensity and practically channeled by the streets. It is also very logical that as cities grow there is a greater quantity of gardens and with that the microclimates are changing. When streets are paved, the drainage systems must be modified for greater conduction capacity from their storm drains.

In any case, the best conditions for life on World are concentrated in regions located preferably at 6,562 feet (2,000 meters) above sea level and surely their multiple twins in other systems with similarity of distance to their suns and with atmosphere or environment that oscillates in climates suitable for life (ideal temperature 68°F (20°C)), since on planets close or very close to their suns the heat is excessive and on the contrary in those furthest from the Sun, the surface temperature is lower than that of World and so by simple proportionality, if we cover the Sun apparently with a one peso coin with arm outstretched, from Jupiter it should be enough with a 10 cent coin. It is good to comment that the Moon can also be covered with one peso coins, which is why during total eclipses the solar corona remains visible where large solar flares can be seen that the Sun gives off and that in many cases cause the black (variable) spots on its surface that are visible through filters and enlarged photographs, or better by projecting it. This coincidence is also worthy of comment (apparently the relative diameters of the Sun and Moon are almost equal).

Returning to what was discussed from Jupiter, the Sun will become smaller and smaller because the planets that orderly follow it, which are Saturn, Uranus and Neptune, will see it smaller and smaller as a large star to small star. Pluto, which was the last to be discovered due to its remoteness (Today discarded as a planet), from it our Sun is confused with the other suns or stars. The rotation of this non-planet with its moon Charon is approximately 2.50 centuries to cover one Plutonian year and its orbit is the most out of phase of the plane of the set of planets, including its axis of rotation is perpendicular to the Sun. Technically it is night all the time since only stars can be seen and the Sun barely from only one of its hemispheres. Its rotation on its axis is not retrograde but not normal either, it is intermediate. Astronomers, due to this great gravitational disorder of this non-planet, were looking for the possibility of a tenth planet but even more out of orbit and since they do not have their own light very difficult to even portray and with the feasibility of being captured by other suns from our same galaxy known as the Milky Way (for its resemblance to a stellar path of concentrated milk in a wide band and similar to a cloud).

4 PLUTO LOST ITS STATUS AS A "PLANET" IN THE SOLAR SYSTEM

A press release said that Pluto, the farthest planet from World, lost its status as such.

About 3000 astronomers and scientists from all over the World met in Prague to decide if Pluto should no longer be considered the tenth planet. By defining for the first time what exactly a planet is, the International Astronomical Union was forced to downgrade it. The main issue was Pluto's status, which is clearly very different from that of Jupiter, Saturn, Uranus and Neptune. Moreover, Pluto's status was debated for decades.

Pluto Unleashed Scientific Fury.- The decision to strip Pluto of its status generated a heated reaction among some members of the scientific community. Experts approved a planet definition that demoted Pluto to a lesser category.

But the chief scientist in charge of NASA's robotic mission to Pluto harshly criticized the decision and described it as "shameful," while the chairman of the committee responsible for making the decision stated that the vote had been "influenced."

The vote took place during the general assembly of the International Astronomical Union. Only 424 astronomers who remained until the last day of the meeting participated in the survey.

The initial proposal to add three new planets to the solar system - the asteroid Ceres, Charon (Pluto's moon) and the distant world known as 2003 UB313 - faced considerable opposition. What followed was a heated debate, and to date there are only 8 planets.

Another comment is that Pluto, which was known as the ninth planet due to the high eccentricity of its orbit, is currently closer to the Sun than Neptune and will move away and later approach again as its trajectory continues. Almost all planets (meaning wanderers) orbit virtually on a plane, not so with Pluto and its moon Charon, which have an orbit that deviates from the previous plane.

All orbits in the solar system are elliptical, meaning they have two focal points, one of which coincides with the Sun and the other must accommodate some gravitational resultant (lofty words because some researchers dare to mention antiplanets). Perhaps the other focus is occupied by pure invisible dark energy (vibrations).

Here I am giving free rein to my imagination, but it is necessary to do so at the risk of being criticized favorably or unfavorably. Otherwise I would be wasting my time translating these thoughts into this publication. In times of inquisition, I would also have had to say the opposite.

I try to avoid science fiction and of course along with it astrology, and instead focus on what seems real to me, which certainly leaves room for much imagination. Before leaving our solar system to get closer priority to the stars in our galaxy, we also have to talk about comets, which as we know describe hyperbolic orbits, meaning their arms open up and consequently their return to the Sun would not be possible (the arms are not even parallel as occurs with parabolas whose arms join at infinity). Even so, after long years, the Star of Bethlehem, the rather large Halley's Comet, **is seen every 76 years**; which from my perspective has the arms of its trajectories asymptotic to the arms of two other trajectories that the same comet has, and it has to orbit at least three stars (suns) for its return to occur, since these asymptotes can form a triangle, with each of said Suns being precisely one of the foci. The tails of comets contain a lot of water and apparently transport seeds of fruits and plants from one system to another.

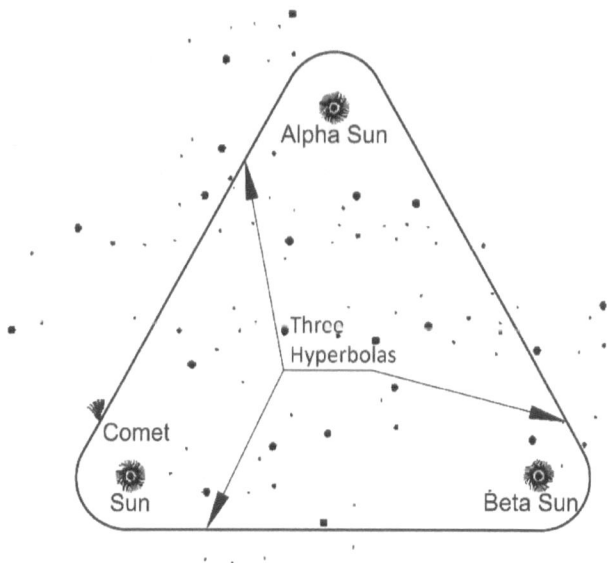

Hyperbolic trajectories of the comet

The World is a huge electromagnet with magnetic lines that go from one pole to the other and generate an infinity of electromagnetic fields, and surely all of this has to do with the rotational movements of the planets and huge gravitational forces. It is interesting to remember that in the universe everything, absolutely everything is in relative or own movement, and it should even be remembered that each of the atoms is equivalent to a tiny planetary system that is in movement, although if we see a metal piece furniture or a glass plate, these are seemingly fixed objects but their particles and atoms move because their electrons are in permanent rotation. Apparently, the water in the seas and the air are the materials that most noticeably move on the planet's surface since their molecules are in a glass of water for a while, running through the sewers at other times, evaporating and forming clouds in which they travel faster before precipitating back into the seas and continuing the hydrological cycle. In short, water goes from Paris to New York (or any other city) and links bodies and beings that retain it momentarily but despite this, as whole it moves with the World around the Sun, and this one in the Milky Way next to the Andromeda constellation and other eternal journeys, and for sure everything is in permanent movement. We should comment here that water is also retained for long periods of time (relativism) and this occurs at the poles and mountains with eternal snow.

This shows that electromagnetic lines play some role, at least in the electric charges of clouds that discharge in storms with thunder and lightning, and in parallel in the World's case, orient us by pointing to the magnetic north with compasses, generating crystallization of minerals in cubic or generally polyhedral forms.

Regarding electromagnets and as a way to take a break from the stars and show a little of the apparently advanced state of our planet, the Railway Technical Research Institute of Japan has built and tested a vehicle called Maglev (magnetic levitation) which when it reaches 100 km/hr (62 mph) begins to levitate about 10 cm (4 inches) above its rails, retracting its wheels like airplanes do to move through the air, being able to reach 500 km/hr (310 mph). So at some point in this millennium, the 21st century, magnetic levitation will be the main technology for medium and long distance ground transportation as oil reserves are depleted. Other countries like Germany have also manufactured these vehicles that are in operation, making mass transit.

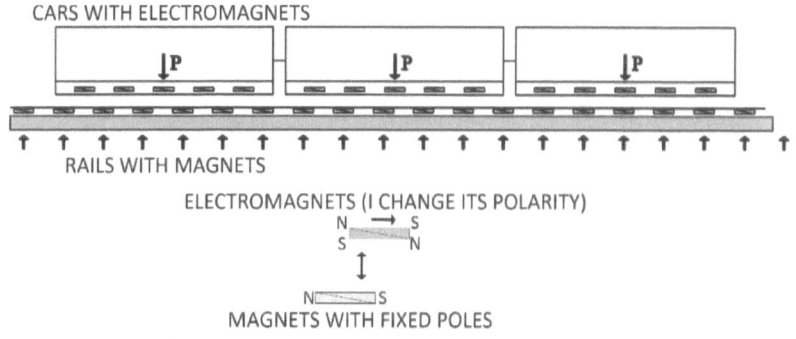

Levmag (Magnetic levitator)

In English it would be called *Levmag* (Magnetic levitator) contains extremely powerful cadmium electromagnets that do not require an electrical excitation source at all. The electromagnets, once energized, are able to achieve very intense magnetic flux densities even in the absence of a power supply since their windings are made of zero resistance superconducting material.

The electromagnetic coils present the difficulty that they must be kept below -269°C (-452°F) to retain their superconducting capacity, which is achieved by housing them in containers with special insulation, constantly supplied with high pressure helium (the liquefied compressed helium is loaded by the vehicle and must be refrigerated all the time), with the cooling compressor being the only thing that consumes energy on board. In addition, these magnets are being improved with superconductors and will reduce the previous temperature to -120°C (-184°F) by changing from helium to nitrogen.

The vehicle's super magnets send the flux both vertically (opposite to World's gravity) and horizontally through the floor and walls of the vehicle. The horizontal is for propulsion and the vertical for levitation. The wall magnets are switchable and change their polarity, in addition there are thousands along the track, while the vehicles have constant polarity and are always ON. A circuit switches each magnet's current from off to on, and controls the polarity of the current so it is able to alternate the polarity of each magnet individually, changing north to south. That is, the north-south attraction causes displacement, while the repulsion also pushes forward.

In the floor there are smaller super coils compared to the vehicle's magnets and these induce repulsion from the vehicle's magnets because they are set to the same polarity while passing through neutral, meaning south to south repels and at the same time in another magnet north to north also repels. Levitation equilibrium is achieved when this force equals the vehicle's weight. (They occupy a continuous concrete channel that is a continuous track crossed by bridges or overpasses and pedestrian walkways).

In Germany the first magnetic levitation train "Trans rapid" has also already been inaugurated and of course by now many countries are working on it.

The research in our time in the world makes one think that this Levmag application must also be under investigation for use in airports and particularly on aircraft carriers since both takeoff and landing occupying the shortest runway length is of interest. This explanation is made to show how advanced the world is and for sure several other planets, some more some less, but always within the limits of each planetary system, and if they have made or established interplanetary contacts, they will have achieved it among themselves (close to each other, but very far from us). In my opinion, the world has a long way to go to achieve it, since, if extraterrestrial visits took place, the logical and natural thing would be for contacts to be established because given the length of these journeys no matter how much the speed of light had been exceeded to make them, it would show that the visitors have taken too much time and logically have aged and consumed all their food supplies. The most that will happen is the establishment of a radio or even television contact which would be wonderful because the language barrier is resolved with signs, but that will also have to take its prolonged time to receive the images responding to the acceptance or denial or confirmation of having established contact, clarifying here again that we do not intend to venture into the field of science fiction, in which case we would abstract from time and assume that contacts are faster.

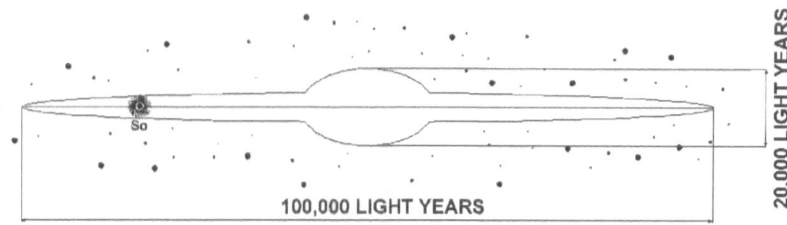

Dimensions of our galaxy (Milky Way)

Research in the field of electronics is quite advanced in the World, but it does not even date back a millennium which represents nothing on the astronomical time scale, plus what is being enjoyed from it is within a hundred years.

Our solar system is located in the galaxy known as the Milky Way and as we know, the shape of this galaxy has been described quite accurately thanks to the fact that the location of our Sun is far from its very congested center of suns (you can't see the forest because of the first trees blocking it). Consequently, from our sufficiently external location, all the bright points (suns, other galaxies or small satellite galaxies of ours) have been recorded and this is how both the top view and profile have been made (with dimensions in thousands of light years). It's not like a ship has been sent to portray our galaxy because for that we would have to hire cameramen from other galaxies to send us photos from an almost fixed site given the enormous distances, that is, contacts with planets from other galaxies would have to be made, and from perpendicular and very distant positions to achieve the described views, apart from getting said photographers. The distances are so great that we would have to talk in terms of thousands of light years for the round-trip message.

SUN

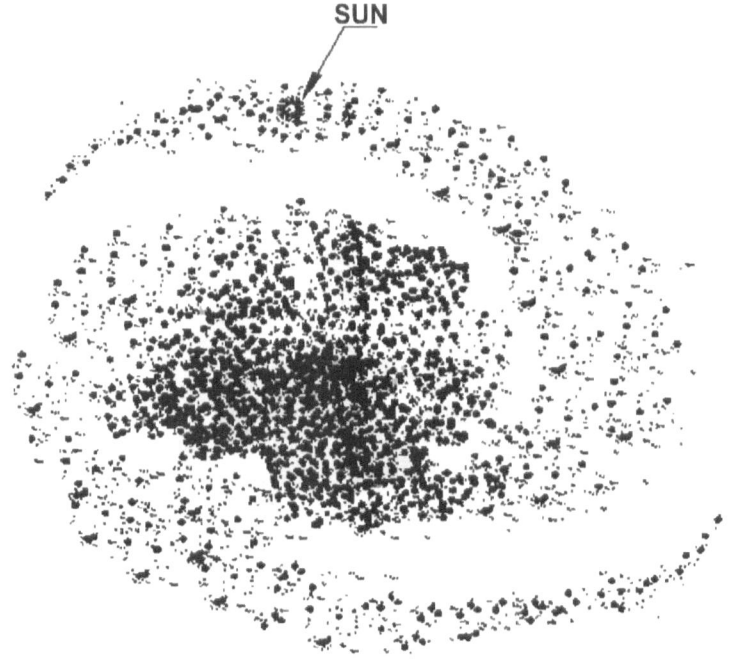

Plan View of the Milky Way

Surely with the resource of radio photography or fax, it will be more comfortable to send books as long as they can read them containing hieroglyphics or graphics to interpret several signs and signals at once because the idea is to chat to exchange knowledge and thus show if they wear clothes, woven or animal skins, or not, and what they eat, if they use water as a basic element for life, in other words, how similar they are to us, as long as their evolution in electronics is similar to our evolution in comparable eras. These contacts have probably already been established among several other exoplanets.

Without dreaming too much, suddenly on a distant planet that surely is not the one we talked about before, some or several of our television programs may have already been captured, which would require a lot of coincidence with the techniques, that would have taken their time to get there. That is, maybe the first broadcasts made in our cities even before satellite transmission. In any case, what they would be trying to do is respond and for that their first problem is addressing the planet World which will have moved a great deal within its galactic ensemble, or perhaps they have already managed to send a response message, but it is taking the same time or similar to the time it took for our signal to move away. The directionality towards broadcast directions also suggests that these dissipate which is why there are radio telescopes consisting of several antennas even arranged at large distances (on our tiny planet) but trying to point to a specific place and with them, images of very distant galaxies have been obtained and by comparison with local data, the shape of our Milky Way has been established. Its center is plagued with suns or stars giving the appearance that the satellites of each one, which are very hot planets or directly also suns, and the same must happen in the other thousands or rather millions of galaxies dispersed in the vastness of the Universe.

The ideal would be to make contact with beings (who are certainly not Martians or products of humans' distorted imagination like the E.T. extraterrestrials) from planets in other galaxies, because that way, the least we could achieve is that they take a photograph of our galaxy and send it to us via facsimile, the only earthly way to broadly know ours, because they already have ones of theirs at NASA. The thing would also be to send them images of several of the exoplanets in their supposed galaxy.

It is necessary to insist on the extremely expensive space journeys at the moment just in our solar system and no further, but the reality is very discouraging because it is not about increasing the size of the ships or the number of astronauts together on a trip. It is also about including in the trip the reestablishment of environmental conditions for prolonged survival (air, water) and food which so far we know is practically concentrated, occupying very little space. But in the long run we will have to consider establishing gardens with crops and irrigation, more fertilizer to have harvests throughout the trip or even better at intermediate stations (artificial satellites), without forgetting that upon moving away from the Sun we will have to use the heat from the other suns that not only heat but also provide solar power. Some domestic animals too, and possibly breeding them because none of World's animals have been imported from space. NASA has surely already considered these things. It would be good if they tell us in more detail how they thought to solve these problems, which make me think about the year 1492 when Christopher Columbus asked the Queen of Spain to send him on the caravels to discover the American continent which was probably already discovered by the ancestors of the original inhabitants of America. Or if human life arose on this side, it would have moved in reverse, but in any case, the biggest worries must have been food and non-salted water because even when going on an excursion a good space in the vehicle is taken up with food, accounting for the time the trip will take.

In modern space travel, the fact that it is daylight all the time must be a concern, with a black sky in the background and millions of bright stars across the sky, because eventually in ecliptic situations one will enjoy the real night while the eclipse lasts. That is, artificial light is surely used to manage this so that the astronauts sleep the equivalent of night on the trip and then when waking up, bathe, shower with pressurized water perhaps upside down or standing in zero gravity. Surely they use soap and sponge. The water will try to clump together but drainage must be by absorption and then recycling so that once clean, they move on to having an appropriate breakfast given the circumstances. Their personal needs will become space waste which surely disintegrates, or they will return it to World. The other alternative is the possibility of recycling and transforming all material on board, precisely so that food, oxygen and water do not run out, that is, with machinery that makes these transformations, at least of the basic materials for the human beings' subsistence.

In current news reports it is said that ice has been found on Mars. I think its existence has been known for a long time, including that its location varies according to its climatic seasons, alternating between its winter and summer there is noticeable variation at its poles. But apparently it is carbon dioxide ice and not water, which should be verified by the presence of human beings in the not too distant future because I think Mars has already been visited quite successfully. To date there was a vehicle with solar power whose movements and removal of soil or dust or Martian stones and/or the presence of other gases in its atmosphere were unknown. Much earlier there is another one on the Moon, it is also unknown whether it continues to move, if it is sending information or what it is doing.

On our planet, the energy source based on oil and its derivatives is causing damage to the environment since the combustion gases are clogging the atmosphere and on the other hand, plastics, which are other derivatives and whose deterioration is very long-term, unfortunately end up coating plant surfaces both in river deltas and at sea floors especially, with water dragging plastics that reach the seas where plankton, the fish's food, disappears. The amount of plastic currently produced would seem to end up coating or plasticizing the World. In addition, there are rubber tires and other synthetics, and worst when they are incinerated producing black smoke due to their carbon content which poisons the atmosphere.

That is why attempts are being made to replace them with solar, wind and hydro power, probably electric vehicles and solar power are the least polluting.

With modern technology, blurry old photos obtained by astronomers are being eliminated or replaced, refining the view of extremely distant objects with very little light because the suns around which their planets are found generate so much light that it is almost impossible to see their satellites. Also, when stars' light penetrates World's atmosphere, they twinkle due to temperature differences in the gaseous layers covering our planet's surface. To avoid this, optical technology must be applied to eliminate twinkling. The most practical thing seemed to be going out into the interplanetary space away from World's atmosphere as has been done with the Hubble telescope. In any case, advanced anti-twinkling optical equipment adaptable technology has been installed in the main observatories, managing exceptionally clear images and even better than Hubble's. Amazing views of galaxy centers have been taken.

Independently of the above, NASA has many space projects including *the International Space Station, the Cassini spacecraft in route to Saturn, the small Near probe which successfully landed on the asteroid Eros. The Curiosity rover traveled for about 9 months until on August 6, 2012 it successfully landed on the surface of Mars always aided by orbiting Mars spacecraft that permanently sent signals to World. The transmitted images took about 14 minutes to arrive after the event.*

Now it is worth commenting on NASA's most important mission for several years: **the World-like Planet Finder**, in charge of locating a planet similar to World far away at distances less than 65 light-years, orbiting a star like our Sun at the right distance for liquid surface water and thus a good chance for life.

At the time of writing all of the above, confirmation has just come through the media and given its relation to the subject at hand, we transcribe it almost verbatim as it came in publications through different communication channels.

In June 2002, American scientists announced the discovery of a planetary system similar to the solar system, as well as the detection of 13 new planets, bringing the number of known extrasolar bodies to over 90 by then. As time goes by these discoveries are growing very quickly so it is not right to give exact current information on the number already discovered.

According to reports, after 15 years of research in this regard at California's Lick Observatory, around a star very similar to our Sun a gaseous planet similar to Jupiter was found, orbiting at almost the same distance as Jupiter's subsystem in the solar system.

On the same star 55 Cancri, in the constellation Cancer, another planet had already been previously detected in 1996, a gas giant slightly smaller than Jupiter that orbits the star every 14.6 days at just one tenth the World-Sun distance, very close to it.

All other extrasolar planets discovered so far orbit closer to the mother star and most in elongated, eccentric orbits, said Geoffrey Marcy, astronomy professor at the University of California, Berkeley.

*"This new planet **orbits at almost the same distance as our Jupiter** orbits around the Sun"*, he said.

According to the researchers also in June 2002, they announced that the planetary system similar to ours could contain a planet the same size as World which would increase the possibilities that it harbors life.

A team of American astronomers has discovered, for the first time, a planetary system similar to our Solar System, representing a big step in studying the possibilities for life on other worlds.

The new system, relatively close to us, is located about 50 light years from World, in the Cancer constellation, and has been named "55 Cancri" after the star at its center.

Using astronomy techniques that measure the light coming from that star, allowing the existence of nearby planets to be inferred, the astronomers discovered that it contains at least three planets similar to ours.

There is even the possibility that the system could contain a planet the same size as World, which would multiply the chances that it harbors life, the researchers said. Because of all this I believe there are extraterrestrials as human as us, hopefully we visit them someday or more comfortably, they visit us, we would give them a big welcome.

Geoffrey Marcy of the University of California, Berkeley, and Paul Butler, two of the most famous "planet hunters," as they are affectionately known in the astronomy world, spent about 15 years making this discovery.

So far, about 80 planets outside the Solar System had been discovered, but all with a size or type of eccentric orbits around their suns, very different from the eight that orbit around our Sun.

"At that time the discovery of a planet was announced which, for the first time, resembles our Jupiter, within a system similar to ours," Marcy said at a press conference held at NASA headquarters in Washington.

His colleague Butler jokingly noted that the new planetary system should be considered "a cousin of ours" because it has many similarities, although also differences.

The three planets discovered so far around 55 Cancri maintain practically circular orbits, making them stable. Two, one the size of Jupiter and the other slightly smaller than Saturn, are very close to the star and a third, slightly larger than Jupiter, orbits about four times the distance that separates Jupiter from the Sun.

"But there is a large space in between, a large region where a planet the size of World could exist and may be stable, according to the calculations that have been made", Geoffrey Marcy pointed out today. Relatives and even our clones must be there.

5 TWISTED MIRRORS SHARPEN SIGHT

To achieve maximum precision, astronomers have combined light from two or more telescopes to create a virtual super telescope, a technique called optical interferometry. They allow astronomers in New York City to observe a cockroach on the Eiffel Tower.

Twinkle little star are words that make astronomers frown. The intermittent brightness of the stars occurs because the light that travels to the World is deflected each time it passes through an atmospheric layer with a different temperature than the previous one. The beautiful spectacle is a nightmare for science: the dancing images look like splashes through the telescope.

There are two solutions and neither is easy. One is to locate the telescope beyond the atmosphere, which is the reason for the Hubble Space Telescope. The other is to use optical technology to eliminate the effects of turbulence. This approach has a big advantage: it can produce clearer views from observatories in the World.

Consumers can already purchase video cameras or binoculars that eliminate image shakes. Using motion-resistant gyroscopes or a set of internal floating lenses, the binoculars produce views that remain stable even when the user is in a moving car. Some video cameras use electronic systems that manipulate the image itself to eliminate these defects. The astronomers' challenge is much greater, since several parts of the image are continuously distorted differently instead of receiving shakes in unison.

After years of development, an anti-flicker technology called adaptive optics is being installed at most major observatories around the World. "Adaptive optics provides images as clear as those from Hubble," said Bruce Macintosh, an astronomer at Lawrence Livermore National Laboratory.

Adaptive optics reflects the primary image on a flexible secondary mirror. Tiny flexible plungers, or actuators, push or pull that mirror quickly, deforming it like a pair of flexible contact lenses, to compensate for every moment of atmospheric turbulence. A sensor analyzes the incident light and coordinates the frenetic movements of the 349 actuators, which adjust hundreds of times per second to keep all parts of the image aligned.

In the most developed version, observers at Lick Observatory on Mount Hamilton use a laser beam to excite a small group of hydrogen atoms high in the atmosphere. This distant yellow flash acts as an artificial reference star, allowing the adaptive optics sensors to accurately measure the atmospheric distortion above the observatory. The Lick prototype has produced exceptional views of the centers of galaxies, and other observatories are already planning to have their own synthetic stars.

In larger telescopes, the adaptive optics system works

only with infrared wavelengths. Infrared rays suffer less atmospheric distortion than visible light, but the giant mirrors intercept such a wide section of turbulent air that their actuators can barely keep up. For this reason, adaptive optics cannot help where astronomers most want it: in clarifying visible light images from the recent super telescopes, whose mirrors are over 14.5 feet (4.5 meters) in diameter. Macintosh thinks it will be "several years" before suitable technology is achieved.

The army uses similar image correction methods to sharpen views taken from space. During the 1980s, spy satellites had a resolution of less than 30 cm, enough to detect a head but not a bald spot. New satellites may be capable of achieving this.

Within a few years, ground-based observatories will be able to surpass space observatories and people with vision problems will be able to enjoy what has long concerned astronomers for some time: the twinkling of the stars.

The new discoveries by these two astronomers have raised to 91 the number of known "**extrasolar planets**" as those not belonging to our system are called (to date, this number continues to increase enormously).

"Out of the more than 100 billion stars in the Milky Way, the constellation where our system is located, there is potential for millions of planetary systems, which opens up enormous possibilities for conditions like those that have given rise to life here," the scientist said.

Alicia Weinberger, an astronomer at the Carnegie Institution and an expert on World's magnetism, considers this finding to be of great significance due to the difficulty of detecting planets outside our system. It should be clarified that everything that is being found at this time occurred at least half a century ago—the time it took for its light to reach our observatories.

The great distance at which they are found makes it totally impossible for their light to be captured from the World, and their existence can only be deduced from the "trembling" that is seen in the light of the star, over which the planet also exerts a gravitational attraction.

The press published the following: Even just a few light years from our location in the Milky Way, a planet similar to ours is so tiny next to the Sun that finding it is like trying to spot a firefly in front of a powerful spotlight. However, NASA began working on the multimillion-dollar project called the **Terrestrial Planet Finder (similar to World)**, with lower than expected costs and timeframes. After just 6 months of 1966, the first few planets found were immense gas balls much larger than Jupiter and hostile to life, **but if large planets exist, why not other worlds with the same animal, plant and mineral kingdoms?**

The best way to find small planets was to move away from conventional telescopes and build an interferometer, consisting of a series of telescopes that harness tremendous power by taking advantage of an optical principle that allows them to resemble a single 15 km (9.3 miles) cross-section mirror, with the largest current telescope having a 10 m (32.8 feet) mirror. What could be captured are planets like Jupiter. The only idea attempted by someone was to use a special filter that blocks the light from the star in question and makes it darker towards its center, that is, to opaque the main star of the planetary system. Infrared and cat's eye deformations were used, which has achieved correcting the Terrestrial Planet Finder, which consists of a single satellite and not four as desired when the idea began, and is located at a stable gravitational level where several stationary satellites are located. With decades in advance of what was planned, new planets very similar to our World have been located.

In NASA laboratories, virtual planets were manufactured.

U.S. biologists, astronomers and computer experts worked together to create planets in order to explore the variety of celestial bodies that could harbor life.

The process of manufacturing new worlds was carried out in a NASA virtual laboratory, which wanted to know what to look for when it launches its next planet detection missions in the coming years.

Through simulations of several planets, NASA's Jet Propulsion Laboratory in Pasadena, California hopes to target the search to those that are habitable.

What Vikky Meadows, project manager, said was "We try to build World-like planets inside a computer. This will help us determine what the vital signs of an extrasolar planet would be once the technology to study them is available".

Around the same time, new results arrived from NASA. Several scientists who used a robot probe have precisely determined the age of the universe - 13.7 billion years old they say it is - and have also calculated when stars began to shine.

Astronomers have been making estimates for decades, but a spacecraft that in 2003 was 1.6 million kilometers (1 million miles) from World was able to look into the past, almost to the beginning of time, to find the answers, NASA researchers said. Stars began to shine just 200 million years after the theoretical Big Bang, the great explosion that marked the birth of the universe, scientists said when announcing the findings of the so-called WMAP mission.

The mission observed the universe when there were still no stars or galaxies, nothing except infinitesimal temperature differences. The WMAP (Wilkinson Microwave Anisotropy Probe) mission analyzed the past, just 380,000 years after the Big Bang, the explosion that according to many astronomers gave birth to the universe. Not even the Hubble Space Telescope can see that far back in time.

Another press headline read: **SIBLING PLANETS?** A nearby star could harbor a planetary system similar to ours.

There was a time when being an Worldling was something unique, a fact that is still preached by some religious sects. In a universe full of stars but devoid of planets, our picturesque solar system seemed to be something special. However, last week the statement said, the club of planetary systems stopped being so exclusive after the announcement by a team of astronomers of a new discovery: a Jupiter-sized planetary body orbiting a not-too-distant star. This suggests that the star may harbor in its bosom a planet with characteristics similar to ours (it contains 4 candidate planets), perhaps two or more with human beings.

As of 2013, at least 17 billion planets of a size similar or close to that of World would be found in the Milky Way according to an estimate that was also published.

Here I want to remember that the time we humans apparently control is so short or minute that we can easily state that Jesus Christ and his apostles or Greek sages are our contemporaries, because humans have been on World for 2.5 billion years. So two and a half millennia are practically nothing, perhaps in the future (another few thousand years) our skulls will be confused with those of the beings that inhabited this planet in that span of time, that is, they could even be confused with those of the beings we mentioned. I am going to mention next some sages who are our contemporaries, for the reason explained above, and trying to make understand the immensity of the time elapsed and the short time they spent alive on this planet, especially great personalities, that is, known to us.

We start about 300 years before Christ approximately and various later dates commenting on a summary of their research during their time on the planet:

Archimedes (287-212 BC) Made known the existence of the center of gravity and thereby the theory of the lever. He established the principle named after him regarding the buoyant force exerted on a body floating in a liquid. He invented the mathematical method of integral calculus.

Leonardo da Vinci (1452-1519) Related static moments to the equilibrium of bodies. He prefigured Galileo's work on uniformly accelerated rectilinear motion.

Galileo (1564-1642) Published the kinematic laws of falling bodies. He discovered the law of inertia which was later formalized by Newton. He observed that forces produce accelerations. He discovered the parallelogram of motions (vector composition) and obtained the trajectory of a projectile. He built the first telescope and with it made profound astronomical discoveries, despite its limited range on the order of current binoculars.

Kepler (1571-1642) Based on Tycho Brahe's observations, he enunciated three empirical laws of planetary motion. He conceived gravity as analogous to magnetic attraction.

Descartes (1596-1650) Enunciated momentum and vis viva which we now know as kinetic energy. Descartes was also a philosopher who relied more on his theoretical conclusions than on experimentation.

Pascal (1623-1662) Applied the principle of virtual velocities to the statics of fluids. The laws of fluid pressure evolved (pressure that is transmitted equally and does not decrease at any point inside a fluid, etc.). He was the first to demonstrate the variation in atmospheric pressure with height.

Huygens (1629–1695) Created the theory of the oscillation center. He invented the pendulum clock and its escapement; by means of pendulum measurements he determined g. He created the ideas of centrifugal force and centripetal acceleration (that is, he deduced an=v2/r for uniform circular motion). He invented the cycloidal pendulum. He established the connection between work and kinetic energy.

Newton (1642–1726) Discovered the law of universal gravitation. He formally enunciated as axioms the laws of motion, which form the basis for mathematically describing the dynamics of a system. He generalized the idea of force; he introduced the concept of mass; he clearly formulated the parallelogram law of forces; he clearly established the law of action and reaction.

Jacobo Bernoulli (1654-1705) Deduced the law of the compound pendulum and the oscillation center from the principle of the lever. He developed a general method (now part of the calculus of variations).

Maupertius (1698–1759) Discovered that the work done to bring a system into equilibrium is a maximum or minimum. In 1747 he enunciated the principle of least action (action = mass x velocity x distance traveled), a vague combination of virtual work and vis viva. Although this contribution was vague, it inspired Euler and Gauss to make some of their best contributions.

Daniel Bernoulli (1700-1782) Discovered the law of conservation of areas, which is a generalization of Kepler's second law of planetary motion. He applied the vis viva principle to fluid motion. He devised a method to determine the output of a fluid through an orifice.

Euler (1707-1783) Published the moment of inertia. He contributed fundamentally to the calculus of variations and applied dynamics methods. He introduced the angles named after him in rigid body dynamics.

Clairaut (1713-1765) Applied potential theory to the equilibrium of fluids. From this point of view, he discussed the shape of the World.

D'Alembert (1717-1783) Applied the principle of virtual work to solving dynamics problems; nowadays his principle is a starting point for establishing the equations of motion.

Lagrange (1736-1813) Systematized both statics and dynamics to reduce science to as formal an operation as possible; his approach being analytical, he opposes Newton's geometric approach. He deduced the famous Lagrangian equations of motion (equivalent to Newton's), which are deduced from energy concepts. Lagrange can be called the founder of analytical mechanics.

Laplace (1749-1827) Applied Newton's concepts to the study of the motions of planets and satellites. His monumental work on celestial mechanics is definitive. He proved the stability of the solar system.

Gauss (1777–1855) Contributed to the theory of equilibrium of liquid surfaces. Orbit calculation methods evolved which are still used today.

Poisson (1781-1840) Discussed the secular stability of planetary orbits. He studied the dynamics of elastic bodies (Poisson's ratio).

Coriolis (1792–1843) Gave his name to the pseudo force $-2m\,(\overline{w} \times \overline{v})$ due to motion confined to a rotating frame of reference. He applied the name work to the product force x distance and denoted vis viva by $1/2\,mv^2$, now known as kinetic energy.

Jacobi (1804–1851) Contributed to the theory of least

action by demonstrating that ∫v ds has a stationary value for the dynamic trajectory, not necessarily a minimum or maximum. He introduced the substitution function, and thereby contributed to the solution of Hamilton's fundamental partial differential equation - the Jacobi equation, so important in quantum mechanics.

Hamilton (1805-1865) Introduced the concept of force functions, negative potential energy. He devised an integral equivalent to D'Alembert's principle. It applies to a dynamic trajectory, which is different from trajectories in Cartesian space (this is Hamilton's principle). He contributed to the Hamiltonian equations of motion, which are now so useful in analytical dynamics and quantum mechanics.

Mach (1838-1916) Exhaustively analyzed the axiomatic and philosophical foundations of dynamics concepts such as mass and force. He criticized and systematized the science of dynamics as it is known today.

Hertz (1857-1894) Criticized the philosophical and axiomatic foundations of Newtonian dynamics. He formulated a forceless system in which only the concepts of time, space and mass were accepted. His main starting point was a combination of the law of inertia and the principle of least constraint.

Poincaré (1854-1912) Created much of the theory of integral invariants and differential equations applicable to celestial mechanics. He introduced topological methods in the study of periodic orbits. He contributed to the study of the n-body problem, the stability of periodic motions, and the stability of rotating fluids.

Einstein (1878-1955) In his special (1905) and general (1915) theories of relativity, Einstein introduced the new concepts of space-time necessary in studying atomic particles moving at high speeds. Variable mass and time are original concepts.

All of the aforementioned, almost two millennia after Christ logically, all and each one of the beings that inhabit the planet, considering the Nobel prizes I insist, turn out to be our contemporaries, because in the timeline they occupy a minimal dimension of the time elapsed.

Continuing with the current discovery of planets, the star of so much hubbub is the one described above, 55 Cancri, a star of similar age and size to our Sun and its planets, located at the modest distance of 41 light years (Previously it was mentioned 50 light years). The news is not new, not at all. Already in 1996 Geoffrey Marcy, professor of astronomy at the University of California and astrophysicist Paul Butler, from the Carnegie Institute of Washington, found another planet also the size of Jupiter that described a very closed orbit around 55 Cancri, only 16 million km from the heat of its king star (a distance even less than that between tiny Mercury and the Sun) In total, astronomers have found more than 90 large planets in orbit around different stars so far; some never leave the reduced sphere of influence of their paternal star; others, on the other hand, prefer to come and go describing irregular orbits, as seems to be the case with another planet detected in the vicinity of 55 Cancri.

Another planet discovered at the same time (clarifying that time passes and this news are left behind), by Marcy Butler and his team is a colossus with very different characteristics: it has a mass between three and a half and five times that of Jupiter (which, in planetary parameters, allows us to consider them practically sibling planets), and orbits at 816 million km or almost like from Jupiter to the Sun). This new planet takes 13 years to travel its orbit, while Jupiter takes almost 12, these figures are surprisingly similar. These differences will be reflected in the clocks that will have different hours for the World and other hours for the corresponding exoplanet.

This extraordinary resemblance to Jupiter is of special interest to astronomers, since one of the factors that made the evolution of organic life on World possible was the close presence of the fifth planet in the solar system. Its enormous gravitational force attracted most of the comets and asteroids that were in the area towards itself, which helped keep the rest of the small planets near our Sun in a relatively safe environment that protected them from colliding with any of the bodies that at that time plowed through the like projectiles. The "open field" created by Jupiter became a kind of orbital backwater in which temperatures could remain relatively moderate, allowing the continuous presence of volatile substances such as water, and thus facilitating the emergence of life. Beyond Cancri 55 and our Sun, many stars also have their own Jupiter-sized planets. However, their orbits usually cross this habitable zone, thus preventing the formation of small World-sized planets. A planet with the characteristics of ours cannot evolve when a steamroller the size of Jupiter passes by - Butler explains.

No one can affirm with **certainty that a replica of our lush and beloved little planet revolves around** Cancri 55. Detecting even a giant planet in the glow of its sun is so difficult that it is actually impossible to see any of these bodies (*astronomers claim it is comparable* **to looking for a firefly next to the beam of light from a powerful reflector**) Therefore, scientists look for small changes in the position of the mother star that seem to indicate that the gravitational force of a larger or smaller planet could be pulling it.

Nevertheless, NASA will launch two new spacecraft in the next 10 years that will describe irregular orbits around the Sun following the World, in order to study the gravitational fluctuations of some stars and obstruct part of their light. Astronomers trust that these projects will make it possible to locate planets that are now hidden. As soon as these ships start operating, 55 Cancri will be one of the first stars studied.

After the transcription described of documents, again and with more momentum we give free rein to our imagination thinking that evidently and with sufficient certainty there are many solar systems with several planets and one or some of them with lives similar to those of the World, that is, vegetable, animal and human, distant from us at minimal distances that would come to be from about 20 light years and much higher within our galaxy and apparently only in its arms because as the stars get closer to the nucleus of the galaxy with 100,000 million stars (suns), life is less likely at least similar to what we know. On the other hand, with certainty the same thing happens in the 100,000 million galaxies many of them similar to ours. And needless to say, about the multiverse.

This description is made to understand that there are millions of probabilities with great certainty of the presence of human beings, plants and animals similar to those of the World, that is, also with World, water, oxygenated atmosphere for breathing, clouds and protective layer from solar or stellar radiation and what is more with temperatures similar to those tolerated by us, and with its good moon to regulate all this.

Given the distances in light years, it is natural to think that the presence of living and thinking beings does not coincide exactly, that is, in some planets they will have dinosaurs or the like and they have not yet been transformed into fuel, as happened in the World and the other extreme that they are much more technologically advanced than we suspect, the energy sources so evolved that perhaps they do not use electrical cables for the transport of the electric current.

In any case, I think that in the World we are in a very technologically advanced era and therefore the findings of the presence similar to our World have just been made and with a history of development very similar to ours. As it is still humanly impossible to think of trips at the speed of light to translate distances into years, and establish visits to get out of doubts, this will be done through the known radio and television communication systems with the hope that in the first Worldly planet to be contacted, it has equipment capable of receiving and interpreting these signals since the languages that these beings speak must be very different from those developed in our World and perhaps with television hieroglyphs will reach us for us to interpret them or they interpret us with advanced electronic languages and respond to us, but if the World that is being photographed is 41 light years away from us, the The answer we will have it in 82 years or our grandchildren and perhaps great-grandchildren who will receive them and again the same for the intercommunication with response from the closest ones. There must be faster energy currents than light when humans discover them, communication contacts will be established in less time.

Consequently, here we make predictions to make similarity relations with the years we have lived and in a very short range such as talking about 10 thousand years before Christ (Christian era) until 3 thousand years later. (Very small space of time in terms of sidereal distances and in the understanding that we are living the beginning of the third millennium and its 21st century of our era).

Beings from some other planets, just living our past should be thinking as it happened anciently in our World, that they live on a flat surface, or theories like those that were developed in the past in which it was believed that the World was carried by some giant or a hemisphere carried by a gigantic turtle or tetrahedral shaped with its 4 faces and later confirmed that it was spherical around which revolved the Sun, the Moon and the stars that included the planets and later ratified the fact that it is the World that revolves around the Sun, the Moon around the World and the 7 remaining planets around the Sun and with presence of their own satellites some of them, on the other hand it is good to remember that in the 20th century, the presence of Pluto and later it was reported that it has a satellite comparable in size to it was discovered and confirmed. For us this story does not have more than 3 millennia that is too little space of time, to date Pluto and its satellite Charon have been excluded from the list of planets.

It is inferred that with absolute certainty there are planets with extreme resemblance to the World in which we live, that is to say with human beings, animals, vegetation and this gives rise to quite a bit to conjecture as follows below.

It is possible given the large number of suns with their planets, and with some of them inhabited they have too much similarity with ours such as very similar days and years in order to imagine that the ages of people also have similar ranges to ours and that between day and At night they can sleep times similar to our night's rest, if they have a moon or moons they will also enjoy its phases and consequently the seas will rise and fall as in the World and also eclipses, but it is too difficult that their clocks are the same as ours. the day is not necessarily divided into 24 hours, with hours of 60 minutes and minutes of 60 seconds. The amount of land occupied and its proportion with the seas must also play a very important role. The way of numbering things will not necessarily be based on the decimal system, it could be in hexadecimal or binary or others.

In these **exoplanets several of them called World twins**, their land if it takes a satellite similar to our Moon, it will be suitable for planting and irrigation especially if they have marked seasons with clouds and seasonal rains, they will do it with natural or artificial channels, that is, they would also have snowy mountains or not, with volcanoes or without them, they will know earthquakes and juvenile and mature rivers, with frozen poles and icebergs in their seas and also tropical areas with temperature variations compatible with those we have. Necessarily to continue making more comments, it is important to place oneself in time as we know in ours just in this age 21st century and beginning of the 3rd millennium, or what is the same end of the 20th century, that is to say that as there are many twin planets of ours, we have to accept their presence not only in our same galaxy but in the same proportion in the 100 billion galaxies, that is insisting, several must be extremely similar to our planet. *It seems that immediate interplanetary communication throughout the multiverse is real, but for us to understand it, thousands of years will pass.*

Many of our twins due to differences of millennia which is very little, compared to millions and which make big differences will be in previous, current and later times than ours, that is to say that some will still not have the presence of living beings, others will have but perhaps not intelligent or very intelligent and others will be living periods such that, for example they still have not crossed their seas in search of the new continent, others will have already overcome supersonic flights or not and in terms of communications, there must be planets that do not suspect the telephone, light, radio or television while others yes they will have already gone through those times, including being in times that we do not even suspect. We if we managed to travel and conquer one of those planets we would become true Mohammed's, as long as we arrive at one more backward than ours.

Another thing is also the Stone Age, Metal Age and others that we have already gone through. Given these time differences between the development of life on our twin planets, I believe we have another reason to show the difficulty they must also have to communicate with us of course thinking that we are neither the most advanced nor the most backward and consequently we believe that the presence of extraterrestrials better extra mundanes is too remote, since if they did it would necessarily have communicated with us that is publicly and not in isolation in the mind of many liars or with imagination in search of personification, besides that these visits could have been before the discovery of America or before Christ or they will just be on the way to arrive when it will be. The possibility that among others twins contacts have already been established between them is not denied, either because they are very close or because of their degree of evolution. It is possible that in the same solar system they have two or more of their inhabited planets, there the contacts would resemble those made to Mars (unfortunately uninhabited for us and only controlled by robots).

We can take the opportunity to ask ourselves if in some of them they have already invented the wheel and engines, or if they still move by horse and in wooden or other more appropriate material boats. At night they will illuminate their streets, and their vehicles will have battery powered lanterns. (This in some of them, because we repeat, the opposite may occur). Do they know about harmful plastics?

What will their roads and/or highways be like, asphalted, plastified, with or without electrical systems for their displacement or by means of magnets and electromagnets, including with good protections on their roads to prevent the violent departure of vehicles or their aeronavigation will be very developed. Much to imagine, because with the aforementioned amount of planets we have to consider the combinations of time, space and movement (without neglecting the law of universal gravitation).

Perhaps they have large viaducts (succession of tunnels and bridges) that allow reducing curves and counter curves by shortening distances including below the waters, for which they must also have good submarines or other means unknown to us.

Another detail that can be talked about a lot is clothing or clothes (men in women's clothing and women in men's clothing with or without hats or perhaps ponchos for everyone, earrings and ornaments, with or without tattoos) because against tastes and colors authors have not written, long-haired men, shaved women or they will have evolved with dyes that not only color their hair but also their skin and so they must have light green people (whites) and dark green (blacks). Will they use toilets or still with pigs in the corral, or modernized ones that do not need to undress or clean themselves personally, will they have showers and swimming pools, swimwear will it be necessary? Will they have notebooks, pencils and books or are they still carving hieroglyphics on stones? In short, the important thing is that women are not drawn without tits, nor men without balls. In conclusion, it can be said that the history of life in the World is described in several of the exoplanets.

We will also talk about housing that does not necessarily have to be the same as ours because many of the materials on our planet have been made with large grinding and baking plants such as cement that allows us kneaded with aggregates of gravel and sands and/or stones, also the use of stucco, brick manufacturing (fired clay), will they have tiles, sheets, ceramics, glass, etc.

O use steel and other metals with optimized strengths, large smelting furnaces are required, will they have them (maybe better).

It is likely that their homes are not very similar to ours, perhaps they use adobe, fired clay or plastics and glass in different ways, such as pre-molded glass and different colors, perhaps they do not suspect mirrors and their applications. We are reaching a point where anything we think will give us too much fabric to cut, for example.

Among the many, thousands or millions of World twins there will be in one or several of them ball games such as soccer, basketball, tennis, bowling, etc., how will they take them to practice, will there be 11 players on each soccer team, will they have two goalkeepers for lack of one and with actual goals, as for the dimensions of their fields, will they be similar to those of our stadiums or perhaps larger with capacity for 100,000 or more fans, will they make local, international or perhaps interplanetary championships if they have already achieved contacts, let's imagine Brazil winning the interplanetary cup of the Milky Way alone and most logically representing the World. The complicated thing would be that the planet in question is a single country, without territorial divisions and with a single ruler. Will they use; leather, plastic, rubber or materials we do not suspect.

Not even suspect of their existence. Also the control of their economy, with or without coins and bills. Bicycles, motorcycles, cars, trucks, cable cars, trains, boats, submarines, airplanes or will they be riding horses or elephants, also pulling carriages.

Moving on to other topics, they will have electric kitchens, gas or refrigerators to preserve their food that they will ingest cold, hot, with or without calories (with or without salt).

Their bottles will be glass or plastic or other material, such as their soft drinks with or without alcohol, if so we must have good drinking countries, or dancers for which music is needed either on record player (old vinyl record player), recorder or modern floppy disks called compact discs, perhaps they play music with equipment that we do not even suspect, and with current computing the removable memories that capture too much amount of musical compositions, but what arises next is, what will their musical instruments be like (drum, piano, violin, guitar, cornet, flute, panpipe, etc.) here they must be very different and with sounds of course unsuspected for us, because their musical scales must be very different or they must not have conceived them at all, all this having always in mind that it is about many inhabited planets and in different states of evolution, on the other hand they will know candles, kerosene lamps, etc.

In short, it will be our future generations in the short or very long term that will be able to see, hear, even smell what happens in the hereafter that we are trying to reach vertiginously, but to tell the truth we are in diapers, because one speaks of thousands or millions of everything.

They will marry or join in pairs or they will do it in the manner of harems (set of women who live under the dependence of a chief), they will take children to school or how will they transmit their knowledge to them.

They will bury their dead, incinerate them, sink them into the sea or send them into space, since they must live a time equal or similar to ours, I do not believe they are more long-lived, but it may be that yes or the opposite.

They will raise babies in cribs and breastfeed them, always imagining that thinking beings are humans like us and with the same characteristics, that is two feet, two arms with hands of five fingers and nails, head with or without hair, with two eyes , two ears, nose with two holes (big noses or not) but as on our planet they must have variety of races, light and dark with straight or curly hair, or with round or almond shaped eyes of various colors or like Japanese elongated if so, there may be racism or not, because they overcame it. Also diversity of races.

They will imprison the mobsters, or liquidate them, drowning them, hanging them, electrocuting them or they will proceed in a strange way for our planet. Will they have prisons or will they deport them to islands? For short, prolonged or definitive times.

They must extract the mineral from the mountains with drilling systems and with explosives activated remotely and with miners who extract it in suitable work vehicles, or they will do it with robotized machinery to then pass it on to purification and smelting plants.

Their food, with breakfast, lunch, tea and dinner or they will do it only once a day, in any case, they need a good sanitary system for the automatic evacuation of waste.

As they will throw away garbage that in some cases they will recycle and in others they will pile up in suitable sites or with aqueducts to transport to landfills or by means of garbage trucks.

As can be seen, the number of questions to space is endless and they will not be clarified in their entirety, in a long time or never (many millennia). In addition, there being thousands or millions the possibility of the existence of inhabited exoplanets, everything commented may have just occurred in many of them, the same, it has already occurred or is about to occur. It is a combination of many elements with many possibilities and alternatives.

Any of the twin planets (several among them) would likely fit well with what we are discussing, and not necessarily all of them. Consequently, we will continue with random ideas, such as, they will make oil, watercolor, or other technique paintings, or they will already have gone through photographic cameras with development and copying or cameras with floppy disks to transfer to computers and thus print touching them up or not, and even deforming images and also superimposing them. Or they will have much more evolved techniques to store images in very reduced spaces and with high resolution. As for film cameras, the same will happen, and most likely they film movies, soap operas or images for their news in three-dimensional systems or not. Some will just be making cave paintings.

On some planets similar to ours they will be divided into countries with rulers and laws, they will have patriotic symbols such as the shield, flag, and also identity documents to be able to move from one country to another and they will do it by air, land or sea. As their airplanes will be, perhaps with flapping wings like birds, or ships with cetacean tails, other vehicles are submarines that will not necessarily be the same as ours and the same with automobiles with legs as agile as those of some animals.

To heroes and important or prominent characters, they will pay tributes to them and make monuments in their squares, the same they will have temples, theaters, meeting halls, etc. One can mention the possibility that the average size of human beings, animals and plants is greater than that of the inhabitants of the World or perhaps smaller in proportion to their planetary diameters or not necessarily and so it would also be that both their cities or dwellings are larger or smaller in size. Some will believe in one God or in several and the same will preach an infinity of religions, everything that is mentioned also with an infinity of solutions because there are millions of planets inhabited by human beings and other species.

Clarifying that not necessarily on all exoplanets, but on some of our twins, as far as sweets are concerned, they will have to know from what fruits they get sugar and how they process it or they do not know it. They will eat dough based on finely ground wheat flour and more properly baked bread in ovens. They will make chocolates based on cocoa, they will know ice cream, how their refrigerators will be or maybe they don't even know them and they go to the ice from their mountains, they will know soft drinks and also alcoholic beverages. They will eat beef, sheep and pork meat, as well as chicken and fish meat or they will act like the Chinese who eat everything that moves, or finally they will be vegetarians or they will eat roots, leaves and tree trunks and for that they will use common salt (Sodium Chloride) or another type of salt, the same with tea, coffee, mate infusions, etc., with or without sugar that is extracted from cane or equivalents to stevia and one cannot fail to mention some vices such as cigarette or cigar consumption and other very serious

80

ones such as drug use such as cocaine, marijuana and others such as candy, chocolates and what comes to our imagination.

All this was mentioned to show that there are marked differences in any case, but the bodies of both humans and animals must be similar on all twin planets, stressing that these issues are in general and accurate in several cases and not necessarily in all. Flat nosed, parrots, hairy, bald, big eared or with small ears, eyed or with reduced or slit eyes.

The following publication, we insert it below due to the chronology with which it appeared, also during the preparation of this work that does not even need chapters. The title was:

THE FATHER OF ALL?. - This chimpanzee simile shook the branches of the trees in Central Africa 7 million years ago. Today he skips the family tree. The story of Adam and Eve becomes a myth to be reconsidered by the church even more knowing that planets in the multiverse are in infinite quantities.

When he was alive he may have looked like a chimpanzee. He roamed the forests and grasslands on the shores of lakes In search of food, accompanied by other members of his own species, always on guard against the lurking of pythons, crocodiles and saber-toothed tigers. He lived in the jungle with other monkeys, and like them he climbed trees. But however similar he was to a chimpanzee, *he walked upright*, which apes only do for short stretches.

However, after death, this same creature has shaken the world of science. After eight exhausting years enduring the stifling winds of Central Africa, an international group of researchers discovered what could be *the most extraordinary fossil ever remembered*. In an article published not long ago in the scientific journal Nature, they reported the discovery of a well-preserved skull of this hominoid, who was almost certainly male. According to Michel Brunet, a paleontologist at the

University of Poitiers and director of the international group, it is not a monkey, but *a hominid*, a member of the subdivision of the primate family whose only living descendant is **modern man**. And that has left scientists speechless for various reasons.

To begin, the fossil is almost 7 million years old, one million years older than the oldest known hominid so far. It almost certainly lived at the crucial prehistoric moment when hominids split in the evolutionary process from chimpanzees, our closest relatives. But even more surprising is that the remains were not discovered in the Rift Valley in Eastern Africa, where our most remote origins have been exhumed over the last three decades, but 2400 Km (1490 miles) to the west, in the Sahel sub-Saharan region of Chad. This forces us to rethink where humankind's true cradle was. In addition, the creature's habitat could well have been a wooded area, which calls into question the already widespread notion that our most remote ancestors originated on an arid savanna. And even more remarkable is the skull itself: the creature, formally known as Sahelanthropus tchadensis ("Sahel hominid from Chad") and informally as Toumaï (which in the Chadian Goran language means "hope of life"), had both hominid features, and particularly the face features, appear much more modern than one would expect at such an early evolutionary stage. In a comment attached to the article published in Nature, paleontologist Bernard Wood from George Washington University states that it is "A hominid less than a third of its true geological age." Paleontologists are already rushing to assimilate the consequences of such a surprising finding. It could be the missing piece to complete the evolutionary process, or family tree, whose latest expression is the current human race. But some argue that the new fossil represents something much more revolutionary that would undermine the idea of the family tree, replacing it with something akin to a dense and thorny shrub. Many scientists now think that the origin of humans was not due to an orderly succession of increasingly advanced ancestors, as conventional hominid family trees suggest.

Apparently it was in reality an evolutionary battle, which was fought at every moment of prehistory, where several species of hominids struggled for the survival of their own genetic stock. Whatever the answer, biological anthropologist Daniel Lieberman of Harvard University believes that "*this is the most important paleontological discovery of the last 100 years.*" All these stories they must be experiencing or have already experienced far from us or they will have to know them again.

Given the extremely hostile environment of the Toros-Menalla region, where excavations were carried out, it is a miracle that this fossil could be found. This desert heat site dotted with dunes lies in the middle of the Djurab desert, four days by road from N'Djamena, the capital of Chad. There are rusty armored vehicles everywhere, remains of 30 years of civil war. Windstorms and sand abound, including scientists, who sometimes had to stay inside their tents for days on end. "They are the most inhospitable working conditions one can imagine," explains paleontologist David Pilbeam of Harvard University, who has worked with Brunet. This makes us think that in many exoplanets they must or must not be at war.

Many spectacular fossils of human ancestors had already been discovered in the Rift Valley, a gigantic fissure extending from the Red Sea to Ethiopia, Kenya, Uganda and Tanzania, but Brunet had good reasons to think that Chad could also be fertile ground for excavations. . In the Rift Valley, occasional rains sometimes erode hillsides, bringing fossils to the surface. Brunet and his collaborators knew that the constant storms of Djurab produce a similar erosive effect. Made up of 40 scientists from 10 countries, Brunet's Franco-Chadian paleontological mission began excavations in 1904.

It was a job of several years, but finally it became clear that Brunet's reasoning had been correct. The group has discovered more than 10,000 fossils of all kinds of animals in the region, and seven years ago they exhumed a 3.5 million year old hominid jawbone. Finally, a student named Ahounta Djimdoumalbaye discovered the amazing remains.

The skull was somewhat flattened and some of its details eroded by the sand, but it was still essentially complete. In a scientific activity where a well-preserved jawbone is already a treasure in itself, this fossil is almost miraculous. Over the next seven months the team discovered remains of at least five other individuals of the same species, including two lower jaw fragments and three teeth. The group did not find any foot or leg bones that would reveal the creature's manner of walking, but **its skull is unequivocally similar to that of the upright-walking hominids**. The researchers would have liked to find these remains between layers of volcanic ash containing potassium and argon, since the date can be precisely determined through analysis of radioactive decay, but in this case the geology of Toros-Menalla refused to cooperate. However, the scientists found an alternative method whose effectiveness is nearly the same; the site abounds in fossils of many other primitive animals, a total of 42 species of fish, crocodiles, rodents, elephants, giraffes and aardvarks, among others. Many were identical to specimens found in other places and already dated very precisely through radioactive analysis. The group led by Vignaud determined that **the age of the skull is between 6 and 7 million years old**.

This places Sahelanthropus tchadensis at the very crossroads of evolutionary change, as scientists have long believed that *apes and humans had a common ancestor*. But recent comparisons of the physical differences between primate fossils, and DNA analysis of modern humans and monkeys, suggest the existence of a single simian ancestral that spawned chimpanzees and hominids between 5 and 7 million years ago. This great-grandfather anthropoid certainly climbed trees in African forests. If so, **Sahelanthropus, or Toumaï**, may well have been the first hominid or at least one of the first to initiate the evolutionary process towards **homo sapiens**. Here's a comment, I believe all the animals that we currently have living in the World are proof of evolution over thousands and millions of years and so they are distinguished for example by dog breeds including descendants of the current domesticated jackal and those of the wolf that represent unstable breeds that is, they are aggressive dogs difficult to domesticate. The same thing happens with plants such as cacti, herbs, decorative plants such as the example of Santa Rita (Bougainvillea) with a variety of colors in its flowers (violet, red, pink, white, brown, etc.), with the characteristic that if they are transported to distant sites or they die or their characteristics are modified such as leaf size and other things. Real quinoa apparently only occurs near the Uyuni salt flat.

But in the very controversial field of human paleontology, *"may well have been"* is a term that lends itself to very heated discussions. It is no surprise that the skull and brain of Toumaï are no larger than a chimpanzee's, as our characteristically large brain did not evolve until just 2 million years ago, much later than "Lucy". But it has become clear that ***Toumaï was not a chimpanzee***, as his face is small with a massive brow. The mouth and jaw protrude less than those of apes, and its canine teeth are relatively small.

However, the place that Toumaï probably occupies in the evolutionary process lends itself to much discussion. Several prominent paleontologists, such as Bernard Wood, Ian Tattersall of the American Museum of Natural History and Chris Striger of the London Museum of Natural History, consider it to be a bewilderingly modern face. Even more advanced than that of "**Lucy**", the Australopithecus afarensis that lived between 3.6 and 2.9 million years ago, and therefore very different from what could be expected from a hominid of its geological age.

Since the discovery of "Lucy" many other older hominids have been identified as our most remote direct ancestor, including **Ardipithecus ramidus ramidus** (from 4.4 million years ago) and Ardipithecus ramidus kadabba (5, 8 million). But Toumaï is even older, and yes, his appearance is as modern as Wood thinks, possibly "Lucy" and the rest were not really our direct ancestors. They could be truncated branches of our family tree, just as Neanderthals were in more recent times. According to this new theory, it is possible that Sahelanthropus spawned species that have not yet been discovered, and these descendants in turn lead to Homo habilis or Homo rudolfensis, the first representatives of the human genus that emerged 2 million years ago.

If so, this seems to support an evolutionary theory that has been gaining ground in scientific thought. Instead of a family tree made up of a trunk and a few branches, hominid evolution would be more akin to a bush where rival species would abound on all sides. The findings certainly seem to point in that direction. In the last twenty years, anthropologists have discovered more and more ancient hominid species, even hundreds of thousands of years old, *and many of them lived simultaneously*.

Apparently, throughout a large part of our ancestors' history, several hominid species coexisted. All this documentation belongs to the Earth's crust, it is to be assumed that in deeper layers there is much more bone information about our ancestors, as well as animals.

If there were several hominid species coexisting for so long, there is nothing preventing us from thinking that this was the case from the origin. In evolution there are abundant cases where new animals emerge not as a single species, but as a series of similar species that share some but not all physical characteristics. The most famous example is the great diversity of finches that Darwin discovered on the Galapagos Islands. Distinguished by their different diets, beak shapes, arboreal, seed-eating or terrestrial, and the variety of canaries in the rest of the world, all of them descendants of a single finch species.

Several anthropologists think that Brunet's discovery supports the theory that there was also evolutionary diversity in hominids. According to Wood, "**Sahelanthropus** could be the first of a group of monkeys and anthropoids that lived throughout Africa **6 or 7 million years ago.**" In the bush genealogy model, not even a surprisingly modern face would constitute proof that **Toumaï** was a direct ancestor of modern humans, and could be one of so many modern-looking hominids who lived during the same era.

Meanwhile, Brunet continues to excavate Chad's stormy desert. "There is still much to be done," he explains. His group not only looks for more **Sahelanthropus** bones, but also older sediments from between 7.5 and 10 million years old, formed by rocks where ancestral species that spawned both monkeys and humans could appear. Paleontologists usually take months and even years to announce their discoveries, and it is reasonable to assume that Brunet and his team have already discovered something else. At the end of his article in Nature, he states that while **Sahelanthropus** will be crucial to explain the first chapter of our evolutionary history "there may be more surprises." Considering the uproar caused by **Toumaï**, that expression could be guilty of diplomatic.

Here we make a comment and that a skull **7,000,000 years old**, for the inhabitants of the planet are already big words because it can easily turn out that all humans descend from him, that is, he could be *Adam or at least Cain or Abel*. For this reason, among the thousands of bones that the archaeologists collected, *Cain's donkey jaw, with which he killed Abel, would have to be found*, but in reality that is not our objective but as an amenity to compare the presence of man in our planet and the others in which in several of them they may be recent in that creation phase, and what we can say is that there must exist in the universe Toumai clones or beings fulfilling their function, asking ourselves at the same time how many years this character has lived and who was his partner for procreation. The story of Adam and Eve is increasingly confirmed not to be obvious.

In other words, beings similar to Toumai are walking on two feet on some planets, and with them it will not yet be possible to establish contacts except that man manages to reincarnate as he currently is on some planet and verify and transmit what he observes. *7 million years represents having traveled a distance 70 times from one end to the other of the Milky Way*:

The diameter of our Milky Way is 100,000 light years and its maximum thickness is 20,000 light years. Impossible to imagine that an astronaut manages to leave it and take real photographs of ours from another galaxy.

Returning to the curiosities on other inhabited planets one cannot fail to comment that in the World and in the field of medicine, organ transplants such as the heart, kidney and some other have been carried out for no more than 50 years, the comment is surely there are planets in Those in which these operations must have advanced much further and naturally in others they have not yet reached this level, that is, as in the time of the Incas they are making trepanations in the brains or not even that, in any case we can imagine that in some planets they make limb transplants such as hands, arms, feet, legs and something apparently exaggerated for our planet, heads or brains.

As far as remedies are concerned, it is likely that based on plants they work at the level of natural medicine or remedies of very high healing power such as saying that cancer and AIDS are no longer as serious as in the World or they have not even known them.

The problem is that once our planet makes contact with these more advanced ones, the cost will be very high for transportation and the worst the time it will take to ship, we better not even estimate it or dream of taking advantage of the most advanced planets.

It is also worth mentioning the advances in optics, that is, they will know glasses, contact lenses, microscopes, telescopes or long-range sights with great resolution, if so they will have managed to see our planet and they will have sent us some message expressly addressed and that, it will be on the way, or it will not have been captured or interpreted by us. By the way, in our World there are radio telescopes and even the Hubble telescope and ships that continue to provide information in space.

The elderly may or may not have vision difficulties and here we can talk about dentures, that is, dentistry may be outdated for what they will drug or get drunk to extract molars or on the contrary they will make dentures or insert dentures and their treatments will avoid repairs and extractions with or without pain.

In reality, they talk about otolaryngology and so hearing will have been saved or improved in general or they will have equipment to listen to secret conversations from a distance, if they have transmitters at the site, the opposite would be that the elderly of a certain age are very deaf and even very blind.

Among the various planets that rotate on their own axis and around their star, whether in the Milky Way or in any other galaxy, they may not know common salt because their seas could be with non-salted water and consequently over the course of millennia they do not have salt deposits to season their meals or if they have them it has not occurred to them to use this mineral as part of the flavoring of daily food. If they have water, it could have washed the salts from their lands and settled it somewhere.

There must also be archaeologists who are concerned with studying their origins, of plants and animals that must also be found petrified or fossilized from trilobites to mastodons or the traces they left in their past, and similar stories to those of Toumai. With a diversity of geological periods and due to geological faults, they would have given rise to the appearance of old strata superimposed on the most modern ones.

One possibility is that planets are already found without life, that is, that at some time there were all kinds of beings and they had built roads, buildings, but due to some impact of external rocks there was a total collapse in the manner of cataclysm killing every trace of life and that only their traces are found, as a sign that there were great civilizations on that planet. In the world it seems that the tendency of its inhabitants is to make it disappear either with bombs, toxic gases and reducing the protective capacity of the atmosphere.

It is very difficult to talk simultaneously about all exoplanets, which is why we often focus on one of them.

These civilizations must have ingested their food with plates and cutlery similar to our knives, forks and spoons or ladles, perhaps with Chinese sticks, pots, frying pans, etc. all made of what material? In Bolivia there is the aptapi (Aymara expression) which consists of the Altiplano groups spreading some aguayos (multicolored fabrics) on the floor and on it they put a series of normally dry foods such as chuño (freeze-dried potato), hard-boiled eggs, corn, ispis (small salted fish) and charque (sun-dried salted meat), and they eat with their hands, possibly all standing or squatting.

They must have known in some of the possible inhabited planets, cheese, jelly, flans and cooked and sweetened fruit or perhaps not or delicacies that we do not know. Others among them with very good liquors or at least with soft drinks.

Always talking about these inhabited exoplanets of the solar systems in the universe in general, the following possibilities also arise.

There may be habited planet pairs in the system in question, which is certainly not happening in our system, in which trips similar to the moon visited by our astronauts would not be so disappointing because contacts between inhabitants of two planets would be equivalent to the discovery of America with conquerors and conquered or perhaps with cultural exchanges and marriages between couples born on different planets, especially if they do not have notable differences in terms of development time.

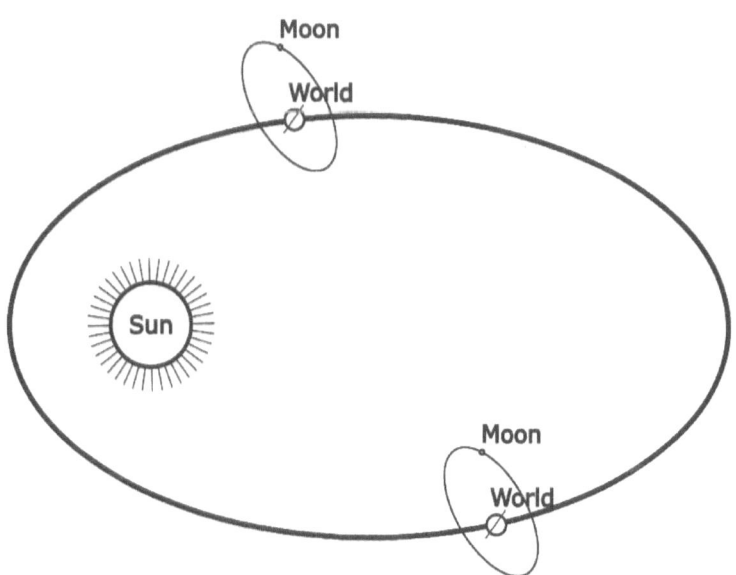

Two in the same orbit (impossible)

These cases of two or more inhabited planets occur with orbits around their sun, and rotation on their own axis or with orbits of planets around another planet, that is with their inhabited moon, in which case communications would be more direct (even with mirrors) but with the duration of their day's variable, except that both have their own rotations on their axes. If these moons have water, atmosphere and protective layer from solar rays and have animal, plant and human lives by rotating around one of them the distances to the sun would vary related to the World from 148,000,000 Km (91,953,919 miles) in +-300,000 Km (186,411 miles) or +-0.625% of thermal variation. Consequently, we can assert that they can have the thermal conditions suitable for life as it happens in our planet, but in these with the advantage of exchange and if not one of the planets without major difficulties, because they can have virgin land and their own seas and beaches and will build planned cities with livable systems, offices, vehicle and pedestrian routes, as well as very well channeled sanitary cables and pipes. They could have very developed wireless systems, in summary, give rise to very modern cities.

Thinking about the existence of two planets in the same orbit is impossible, they would surely collide like billiard balls.

It can also happen that two solar systems with inhabited planets are very close to each other but keeping the minimum distance enough so as not to capture distant planets from their center, as is the case of the former planet Pluto in our system whose distance to the Sun is 5.895 million km or you can talk about a proximity between planets from different systems of the order of 12,000 million km which traveling at a thousandth the speed of light would take 50 days (acceptable) at 300 km/sec to move from one system to another and in parallel positions those of the plane of rotation of both systems and also similar situations occur to those mentioned above, with all their combinations, but for us very distant, this pleasure. Formerly people traveled from South America to Europe in almost a month on ships that were better than many star hotels. In addition, it should be clarified here that inhabited planets are closer to their suns and not on their periphery, which greatly increases the time of these trips.

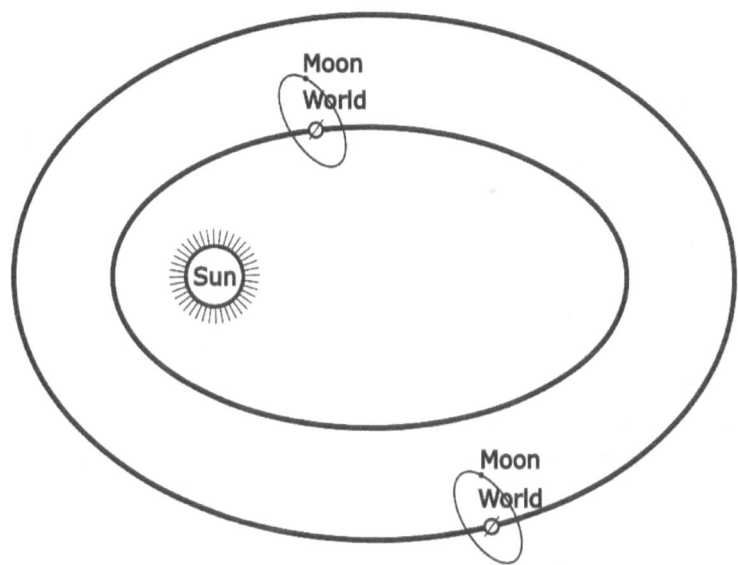

Two in different orbit (Possible)

After everything mentioned, it is to think that comparing with the past in the World, all the wonders of electronics and telecommunications that we are enjoying since the 20th century because we have lived it, something that did not happen in the 19th century where they were just building the first engines, and now in the 21st century we enjoy facsimile, wireless phones the Internet Network, supersonic flights including the spacecraft with which the World has been interconnected so that it seems that it had been reduced in a short time, and with what in the short term there will be a universal language for intercommunications, because it becomes essential, for which the alphabets will have to contain the essential letters to facilitate common reading and writing for all without discrimination of cultural levels, making everything accessible to everyone.

A situation to comment on is the possibility that a planet suitable for life was found in its evolutionary process, that is, it has air, water, animals and vegetation but still no human beings like those on World, but at that time astronauts travelers arrive from the World, they would occupy the place of Adam and Eve if they went in pairs, since they would necessarily have to travel in pairs with the idea of settling on those planets and reproduce as much as possible with several descendants so that they marry and give rise to a new population practically in paradise (without snakes or apples).

In the event that twin planets to the World are really similar to ours and with life and if both equivalent to the World and the Moon are the same size, the occurrence of eclipses on both planets would be similar or not because their axes and trajectories (ecliptic) should also coincide, which, in our case, in which they only occur in certain areas and with maximums of three solar and up to four lunar in one year. It is thought that they will be more frequent because if they have equal or similar dimensions but their axes with less inclination, they will cause partial and total shadows more easily than in our case because the Moon, which is almost a quarter of the World in its diameter, would pass through its ecliptic more often. In other words, when one of them is in eclipse of A, the other will be in eclipse of B and vice versa and probably the influence on the climate in both planets can lead to a predominance of cold.

In the Bolivian town of Huachacalla, Oruro department, we had a total solar eclipse, the same one that we observed from another location (I speak of my family) almost the center of the Uyuni salt flat, on November 3, 1994, that is more in the penumbra area, it was something spectacular because when the shadow of the Moon moved a beautiful curtain came over us surely resembling an aurora borealis, which dragged the projection of the shadow of the lunar hills over the whiteness of the salt flat, at great speed that we did not have time for photographic shots or filming, it was very exciting.

When the Moon is full for the World, it shows the Moon its night side, that is, dark and when the Moon is New, the World will be full for it and the Moon dark for the World and if it is very aligned it will be a solar eclipse, for the first consideration the eclipse It will be the Moon.

Another case to mention is if both with all the conditions of life are proportional in size, one to the World and the other to the Moon of one they would see the other as from the Moon to the World or from the World to the Moon and we can imagine how the phases of the Moon (or World) will look, one similar to what we observe and the other with larger proportions, in both cases one would distinguish with the naked eye, the parts of land, sea and their clouds and at night the lights of the cities if they are like those in the World.

The fact of having confirmed the presence of exoplanets in the universe in general, surely tells us that life similar to ours exists in thousands or millions of cases and with the difficulty of communicating or contacting some of these planets and getting out of curiosity and at least confirm some of the possible cases of human life in some parts of the immense universe to which we will surely travel with our spirits (form of energy) with which they can move very quickly.

It could happen given the number of possibilities that one of them has two satellites and both inhabited, with which eclipses and lunar phases are of great diversity, because one or both may even have retrograde rotation and at what stage of development they will be, what if all this leads us to the certainty of confirming the presence of human beings everywhere and with samples of almost the entire history of our evolution, which would be possible if we manage to contact many of them, but for a long time I consider it very remote. Always convinced that they have the same characteristics as ours and they are not as we are induced to think of Martians with little antennae or funnels or rather monsters, not even animals that we know with minds more evolved or superior to those of our human beings who speak and chat in different languages but intercommunicate and appreciate the beauty, the beauty that is the universe and in our World, with flowers of various kinds, colors and smells, as well as animals with a variety of shapes, colors, emitting various sounds and moving either flying, or running, or walking or in the water like ducks, fish, cetaceans, etc.

All **this is the work of a supreme being known as God**, who has made everything in his own image and likeness, because from the little we know about our planet, it is very beautiful and varied, from the bottom of the sea to the top of its mountains, with territories full of vegetation or without it, with juvenile and mature rivers and possibilities for orientation in any location and many means of transportation and displacement on the surface. Also thinking that the whole range of metalloids and metals is the same everywhere in the universe and what is fundamental will have atmospheric variations, that is, the four seasons of the year. What determines the attachment to life.

In short, practically all known fields of specialty are being touched on this planet and that is why we can say that perhaps overpopulation will make emigration to other inhabitable planets necessary but preferably unoccupied because it is fundamentally understood that emigrants do it because they do not like coexistence with beings who perhaps think differently or do not give them room to develop as merchants or establish themselves as powerful in certain spaces and even manage or lead groups of thinking beings, that is, with desires to be bosses or chieftains or finally lead a solitary or maximum family life in places like jungles or forests or finally try to be prophets in sparsely populated places , but I think that in the World we have everything a human being needs to live well, the important thing is not to take advantage of anything or anyone and respect the rights of others and reciprocally be respected, how beautiful it would be to eliminate padlocks, alarms, doors and even windows confident that no one will take what does not belong to them. As a comment, cars that at some point spread wings and rise like planes are currently being tested.

It is very interesting to relate spatial distances with elapsed time. Thus, Pluto has only completed two orbits around the Sun since the discovery of America, which corresponds to **two Plutonian years**, while the World has completed more than 500 orbits, corresponding to 5 centuries. This can even be compared to the **two millennia** since the **epoch of Jesus Christ**. To date, Pluto has only completed **8 orbits** around the Sun, which is incredible considering our solar system is only one among 100 billion suns or stars. The orbital velocity of this planet is 20 times that of our jet turbines and yet it still takes about 2.50 World centuries to complete a single orbit around the Sun.

Another news article from a few years ago mentioned that American astronomers discovered an enormous planet orbiting our Sun at a distance of 6000 million kilometers (3,730 million miles). It could have become Planet X, but rather Pluto is no longer considered the 9th planet and I believe that with even more reason Planet X would not be accepted. The celestial body, whose diameter is one tenth of World's, is located beyond Pluto in an area of the solar system known as the Kuiper Belt. Scientists described the finding as the greatest made in the solar system since the discovery of Pluto, 72 years ago.

The new non-planet, which was named **Quaoar**, has a diameter of 1280 kilometers (796 miles) and completes its orbit in 288 years. "Its size is like that of all the asteroids combined, it is very large," commented astronomer Michael Brown of the California Institute of Technology in Pasadena. Quaoar has only completed 7 orbits around the Sun since the time of Christ.

Brown and his colleague Chadwick Trujillo detected this new world in photographic images. The new object is called Quaoar, inspired by the creation myth of the Tongva people, who inhabited Los Angeles before colonization.

For the natives, Quaoar was the great force of nature that gave rise to all things. It was provisionally designated **2002LM60**, based on images dating back to 1982.

6 WE LIVE IN THE GOLDEN AGE OF ASTRONOMY

A few years ago, Pierre Gaillard, head of the UNESCO Public Information Office, interviewed Catherine Cesarsky, president of the International Astronomical Union since 2006, under whose initiative the International Year of Astronomy was celebrated in 2009.

Catherine Cesarsky is a research professor at the Atomic Energy Commission (CEA) in France and an associate researcher at the Paris Observatory.

What is astronomy used for? Astronomy is a science through which humankind can try to answer questions that it has always asked itself. Where do we come from? Where are we going? Astronomers try to answer those questions scientifically. We try to understand how the universe works and how it was created, and also to understand how the galaxies that populate it were born, as well as the stars and planets. We also try to find out if there are planets like ours in the universe. In time, we will probably try to find out if they are inhabited by living beings.

What is the current state of astronomy? We are living in the golden age of astronomy. This is largely due to immense advances in technology, which our science uses to the maximum because it is what allows it to move forward, always resorting to the most cutting-edge inventions in the fields of electronics, optics, mechanics, etc.

When we observe a very distant galaxy we see it as it was when it emitted the light that we receive, because the latter takes a long time to reach World. We currently estimate that the age of the universe oscillates around 13.6 billion years. Today we are able to observe galaxies as old as the universe itself.

What advances does astronomy promise for the coming years? We are currently making the first discoveries of galaxies that existed in the early days of the universe. In the future, when we have even more sensitive instruments, we will be able to find out how they were formed and whether they look like galaxies today. We will also be able to study their properties. I find this especially interesting. At this time we are also accurately determining the cosmological parameters, that is, those that govern the expansion of the universe and its creation when the "Big Bang" occurred.

There is a lot of progress to be made in this field. Finally, it has been said that a little over fifteen years ago, extrasolar planets began to be discovered through their indirect effects. Today we already know several thousands. We are getting better and better at locating those that resemble World, and soon we will be in a position to study their characteristics.

Has the astronomer's profession changed with all these advances? Today's astronomer is nothing like astronomers of the past. There are currently two types of astronomical observation: ground-based and space-based. For space observation, instruments take a long time to create, they have to be perfect and there is no room for the slightest error with them. To propose an instrument design and build it is a process that takes about fifteen years in itself. Then the instrument goes into space on board a satellite, and to explore the solar system, sometimes you have to wait another eight to ten years for the probe to reach the chosen place. You need a lot of patience!

Astronomers who observe the universe from the ground use telescopes that are nothing like those of their predecessors. Telescopes now have diameters of 8 to 10 meters (26 to 33 feet), but we are studying the creation of others with diameters of about 30 or 40 meters (98 to 131 feet), and even larger. Astronomers are no longer sitting under a frigid dome, staring at the stars with their eye glued to the telescope and trying their best not to let them out of sight. They work with computers and do everything with remote control systems.

Astronomy professionals are no longer content with visible astronomical observation. From the ground and space we scrutinize the universe from end to end with every possible means, from radio waves to gamma rays, thus obtaining a more complete picture.

New theory about the universe.- A joint team from the United Kingdom and the United States has presented an alternative theory on the evolution of the cosmos.

The proposal presented in Science magazine argues that the universe goes through cycles of "Big Bang" and its counterpart, the so-called "Big Crunch" in which the universe collapses after expansion stops.

Next, a comment on something from the recent past.

What did the International Year of Astronomy provide? The idea for the International Year of Astronomy came from the International Astronomical Union. Fortunately, UNESCO offered its support right away. The goal was for everyone around the world to share astronomers' admiration for the mysteries of the universe. Its wish is that all countries participate at the same time in the activities of the International Year. Many of them have already developed programs for this purpose. We also want the general public to participate as much as possible. At the end of the International Year, all the people on World will have gazed at the sky for at least a moment, or will have read something about the latest astronomical discoveries, or will have reflected a little on the place we humans occupy in the universe.

7 NASA SEEKS TO DETECT PLANETS SIMILAR TO WORLD

NASA (National Aeronautics and Space Administration) launched the Kepler space telescope on its first mission to detect planets similar to World that could harbor life in our galaxy.

The huge Kepler telescope was launched with a Delta II rocket from the Air Force station at Cape Canaveral, and continues to send information to this day.

This was NASA's first mission searching for World-like planets orbiting stars similar to our Sun, at just the right distance from their star and with the proper temperature for the presence of water that could sustain life.

Kepler is a key element in NASA's efforts to discover planets where an environment similar to World's might be found, explained Jon Morse, director of the agency's Astrophysics Division, at a press conference.

The planetary inventory Kepler must compile will be of great importance for understanding the frequency of planets the same size category as World in our Milky Way.

Kepler's discoveries will fundamentally alter humanity's view of itself, Morse said.

Equipped with the largest camera ever launched into space - a 95 megapixel array known as CCDs - the Kepler telescope is capable of detecting faint stars, barely perceptible, with planets detectable as they pass in front of them. Remarkably, this observatory captures faint rays of light to reveal their sources, whether stars, galaxies or others, that would have been impossible to see otherwise.

If Kepler were to point at and observe a small town on World at night from space, it could detect the porch lights if someone walked by in front, according to Kepler project director James Fanson. As a remark, we must forget about nighttime privacy even on our planet.

The Kepler mission, which cost nearly $600 million (€538 million), lasted about three and a half years analyzing more than 100,000 stars similar to the Sun in the region of the Cygnus and Lyra constellations of the Milky Way.

Astrophysicist Alan Boss is convinced that Kepler or the French satellite COROT, which has been in orbit since 2006, will be able to find World-sized planets in the coming years.

It would be absolutely stunning if Kepler or Corot did not find World-like planets, because basically they already are finding them, Boss told the press in February at a science conference in Chicago.

Corot has already discovered the smallest extrasolar planet so far. Just over twice World's diameter, the planet is very close to its star and is very hot, astronomers reported this month.

Smallest extrasolar planet in universe discovered.- The smallest planet known to date, named Gliese 581e and whose recent discovery has just been announced, is the most similar to World found so far, said Gaspare Lo Curto, ESO astronomer in Chile.

It is located 20.5 light years from World and orbits a star around which three other planets have already been discovered. Its surface is rocky and it has no atmosphere because it is too close to its star and temperatures are very high. All these observations are inferred based on computers or electronic processors.

On the other hand, according to the BBC, an international team of astronomers discovered the smallest planet outside the solar system most similar to World ever identified (January 27, 2006). The new planet is 5 times more massive than World and can be found 25,000 light years away in the constellation Ursa Major, orbiting a small red star. The discovery was made using a method called microlensing that can detect distant planets with a mass similar to that of our planet.

Like World, the planet has a rocky core and a probably thin atmosphere. However, its wide orbit and the conditions of its parent star mean it is a cold world. Surface temperatures are calculated to be -220°F (-140°C) which means it is possibly a frozen liquid. Therefore, it could be similar to a larger version of Pluto.

Ring of stars around the Milky Way discovered.- US astronomers discovered a ring of hundreds of millions of stars surrounding the Milky Way that could be remnants of the violent past of our system, announced years ago in Seattle (Washington). This ring of stars could be what was left after a collision between the Milky Way and a smaller galaxy billions of years ago.

The 100 to 500 million stars orbiting the Milky Way were discovered by scientific teams from the Sloan Digital Sky Survey, responsible for making a detailed map of a sector of the sky.

Colored quasars suggest a universe full of smoke.- Take a deep breath: we may live in a universe full of smoke, which diminishes the light from distant objects such as quasars, according to astronomers working on the Sloan Digital Sky Survey 2. In the same way that a sunset appears reddish because light has to travel through more of the World's atmosphere, the light from quasars can appear red because it travels through the dust from the galaxies it encounters along its path. Given the sheer size of astronomical distances, we must not forget that if as a general rule in a light year on each side of an ideal cube containing just one atom of any substance constitutes a sort of smoke.

Quasars are defined as almost red star-like objects with high energy output and are described as stars with an extremely fast velocity, perhaps exceeding the speed of light.

Galaxies contain large amounts of dust, most of it formed in the outer regions of dying stars. The surprise is that we are seeing dust hundreds of thousands of light years outside galaxies in intergalactic space. Using images from the Sloan Digital Sky Survey 2, scientists analyzed 100,000 distant quasars located behind about 20 million galaxies.

Averaging over so many objects allowed them to measure an effect that is too small to see in any individual quasar. There are two theories: One, supernova explosions and strong winds from massive stars push gas and dust out of galaxies. The other is that this process occurs due to radiation pressure on the dust from starlight. Whatever mechanism pushes dust into intergalactic space, its presence there can be a problem.

Ancient Pulsar Still Beating.- The oldest and most isolated pulsar ever seen has been discovered with NASA's Chandra X-ray Observatory.

A pulsar is a neutron star characterized by the emission, at regular short intervals, of very intense radiation.

This old and exotic object has become very active. The pulsar PSR J0108-1431 is about 200 million years old. Among isolated pulsars, those not spinning around a binary system, it is 10 times older than previously discovered pulsars and is an X-ray record holder.

Recent technological advances allow us to have much superior observation capabilities compared to what we had before. For a long time we were only able to study the galaxies near World, what we call the near universe. Now, on the other hand, we have more sensitive telescopes and instruments that allow us to observe much dimmer light sources. Often, this lesser intensity is simply because they are located very far away.

At the time when this book was being finished, news continued to arrive about discoveries such as that of a rocky planet outside the solar system to which World belongs. The exoplanet called CoRoT-7b is 500 light years away and has the same density as World according to the European organization for astronomical research in the southern hemisphere. After months of measurements, experts determined its mass is 5 times greater than World's, clarifying that to date they have only been able to measure the density of 63 of the 330 exoplanets discovered (15 year old data).

It is my hope that this reading *gives the inhabitants of planet World much to think about because we are not alone and our entire past and future history is scattered across a variety of twin planets where there are beings like us, but not necessarily in the same era that we are living in (almost all of known history encompasses just two millennia, which is very little time).* It could be that after our death we reappear on one of these planets similar to World.

Multiple universe theories exist. *Multiverses*, if a grape represents our universe, I believe the grapes of all the clusters in a vineyard combined would still scarcely represent the rest of the universes. This suggests that with absolute certainty identical clones of ourselves exist, but each with its own soul, in unimaginable regions.

8 HIGGS BOSON

The God Particle.- A journalist interviewed a priest on the radio who was a chemist by profession and asked him if the term **"God particle"**, referring to a subatomic particle, more appropriately called the **Higgs boson**, was not an attempt to deny the existence of God. The priest commented that this was not the case, since even the one who coined the name "God particle" for a popular science novel, the 1988 Nobel Prize winner in physics, Leon Lederman, had pointed out that demonstrating the existence of the Higgs boson would help better understand *how God made the inconceivably large multiverse (infinite)*.

The God particle, the scientific discovery of the 21st century, which at the subatomic level was announced as being compatible in my opinion with the soul or spirit which in the case of human beings is located somewhere in the brain and perhaps with movements inside both the right and left sides of the brain, making ambidextrous beings. In my opinion, this boson is fundamentally the element that allows the interconnection of all living beings in the multiverse, whether people, animals or plants.

The complicated thing about the boson is when it is born and when it disappears or will it be eternal like God. To try to resolve doubts, let's ask ourselves if we know that as a result of the union of human couples, the life of a new being arises, and perhaps twins or triplets. Each one with its respective boson which is the driving instrument of the human genome and which could be different if the couple in question were with another different couple, the heirs would have other bosons, I don't think the same ones, one would be the predominant one (unlikely) in generating the new bosons which in turn could come from distant planets, which in this case make us see that the role they play is equivalent to the chip in computers that make matter appear to be controlled at genetic levels, which is why all the inhabitants of the world are different and at best we could be cloned based on the fact that genes contain all the genetic information to do so.

This gives rise to many issues, because it could be that when people die and as the only way to move through space, bosons or spirits come out at perhaps instantaneous speeds through the midst of darkness or dark energy. Since at speeds such as that of light (approximately 186,000 miles/second [300,000 km/sec]) rays of luminous energy pass and break down into the immense range of colors that allows us to see objects painted with chosen and also natural colors such as the green of vegetation and the blue of our atmosphere and consequently of our planet and what is more, the transmissions of images and sound.

In some cases, such as family members who allow us to have an affective intimacy, we know our parents, siblings, cousins, nieces and nephews, uncles, etc. quite well, whom we normally remember as being close to us or away on a trip to another continent, but we do not forget them, even if they have died, and with the evolution of our planet we communicate and chat seeing each other on the internet, though not yet with the deceased. Then on a scale of values come childhood friends, school friends, college friends and work friends who over the course of time we may have forgotten and if someone reminds us of them we retrieve them from memory. And most impressive is that of a partner, particularly if one is truly in love, that being inserts itself so much into oneself that, at the death of one of them, it seems that their spirit has also been inserted into one of the two, like a boson that shares everything with the one we carry inside.

I think the boson fulfills in our body the task of having organized the human genome in such a way that even cloning is possible similar to plant cuttings, because clones can also be generated from these new beings.

But in times like over two thousand years ago, the existence of the soul was known and even the saints had visible halos without the need for the progress or evolution of our days that can allow us to appreciate the halos emitted by our body, especially on hot days.

What happens to the boson of beings who undergo organ transplants? For example, the kidney donor survives perfectly well in most cases, as does the recipient, but in this case it is worth commenting first on the fact that the constituent molecules of human bodies are completely renewed over a period of seven years, including the entire bone structure, which leads one to think that if these beings have exceeded seven years since their operation to date, all their elemental components have become others, so that the donor's kidney ceases to be the one that was originally transplanted. But the matter does not end there because if we ask ourselves what will happen with the cloning of the organ recipient, if the matter of the recipient's kidney is removed the new clone will be a copy of the donor, and if the matter has been extracted from elsewhere it will be like the recipient.

A very unique case is that of grafts of artificial matter such as silicone or platinum, which are grafted into people without rejection, but these materials do not change over time. Even once the body has died and disappeared, these elements do not disappear as they may have been used in cosmetic surgery or orthopedics such as prostheses with screws and bushings to restore bones and their mobility based on joints. Other prostheses are plates in the skull covering the brain again and likewise false teeth.

Repeating, dreaming costs nothing. Research and its valuable results help us, the reports generated are not in vain.

The following graph is intended to show our galaxy or Milky Way (hatched) surrounded by millions of universes, which suggests that as an effect of the Big Bang, our galaxy is expanding and will reach a point perhaps billions of light years away when it is compressed again and other universes expand.

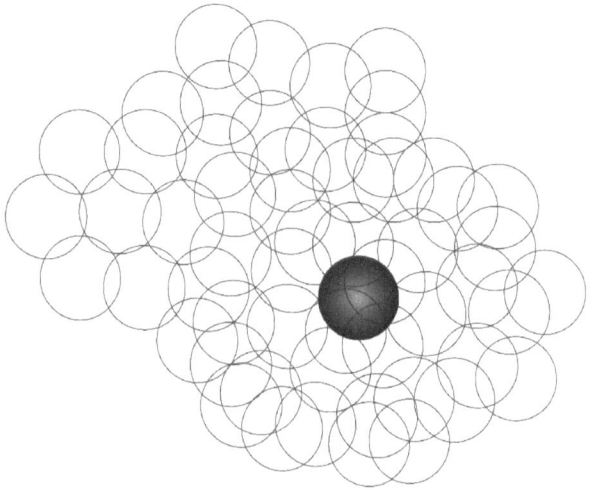

Multiverse (The shaded part is our universe)

We will have to accept that immense entity as God.

They confirm there is a dark energy pushing the universe, although proof is still lacking, predict modification of theory of relativity.

Dark energy, mysterious cosmic force considered fuel behind acceleration of universe, is reality as certain as existence of famous Higgs boson. Finding belongs to group of British, German astronomers.

After two years of research, scientists from University of Portsmouth in UK and LMU Munich concluded probability of existence of dark energy is 99.996%. That is, same level of certainty reached in recent months by Higgs boson, a finding celebrated by global scientific community.

Lead author, Giannantonio Tommaso, noted this work could mean possible modifications to Einstein's general theory of relativity.

In my judgment, future studies of galaxies should provide definitive measurement, either confirming general relativity, including dark energy, or even more intriguing, demanding completely new understanding of how gravity works.The God Particle - A journalist interviewed a priest on the radio who was a chemist by profession and asked him if the term "God particle", referring to a subatomic particle, more appropriately called the Higgs boson, was not an attempt to deny the existence of God. The priest commented that this was not the case, since even the one who coined the name "God particle" for a popular science novel, the 1988 Nobel Prize winner in physics, Leon Lederman, had pointed out that demonstrating the existence of the Higgs boson would help better understand how God made the inconceivably large multiverse (infinite).

9 PREMISES AND REVELATIONS

The basic premise of modern cosmology is that the visible universe of stars, planets and gases is only 4% of the cosmos, and that it actually floats on a sea of unknown material called dark energy.

According to various theories, this dark energy is considered to form 73% of the cosmos and the remaining 27% is formed by the slightly less mysterious dark matter.

The scientists re-examined all the arguments against the dark energy theory and concluded that this is, almost certainly, responsible for the hottest part of the cosmic microwaves.

The scientists re-examined all the arguments against the dark energy theory and concluded that this is, almost certainly, responsible for the hottest part of the cosmic microwaves.

We have systematically addressed all the issues and concluded that none of them can explain what we observe, explained Bob Nichol, a team member.

As background, the following is mentioned:

Clues.- A decade ago, astronomers observed the brightness of distant supernovae and realized that the expansion of the universe appears to be accelerating. This acceleration is attributed to the repulsive force associated with dark energy, which according to current theories is believed to form 73% of the cosmos.

Doubts.- Despite the researchers who made this document, Saul Perlmutter, Brian P. Schmidt and Adam G. Riess, receiving the Nobel Prize in Physics in 2011, the existence of dark energy continues to be a topic of debate among the international scientific community.

The Author

The Spanish-language version of this work was reviewed and updated in June 2014, 21st Century Millennium 3.

.

The English version was translated using AI support in January 2024.

ABOUT THE AUTHOR

He was born in the city of La Paz - Bolivia

Primary and secondary studies at San Calixto School
(1948-1959)

Graduated with a degree in Civil Engineering from the
Universidad Mayor de San Andrés (1960-1966)

Specialization in Prestressed Concrete in Paris - France
(1968-1969)

Structural calculation of numerous buildings in Bolivian territory
(1970 to 2019)

Former professor of Bridge Engineering at the Universidad Mayor
de San Andrés and the Military School of Engineering, with
published work on the topic in its 6th edition.

Stamp collector

He passed away on August 30, 2021